FIRE IN PARADISE

FIRE IN PARADISE

The Yellowstone Fires and the Politics of Environmentalism

MICAH MORRISON

HarperCollins*Publishers*

HarperCollins books may be purchased for educational, business, or sales promotional use. For information please write: Special Markets Department, Harper-Collins Publishers, Inc., 10 East 53rd Street, New York, NY 10022.

FIRST EDITION

Designed by Alma Hochhauser Orenstein

Library of Congress Cataloging-in-Publication Data

Morrison, Micah, 1957–
  Fire in paradise : the Yellowstone fires and politics of environmentalism / Micah Morrison.—1st ed.
      p.   cm.
  Includes bibliographical rereferences and index.
  ISBN 0-06-016303-8
  1. Forest fires—Yellowstone National Park—Prevention and control. 2. Forest fires—Government policy—Yellowstone National Park. 3. Fire ecology—Yellowstone National Park. 4. Environmental policy—Yellowstone National Park. 5. Yellowstone National Park.
  I. Title.
SD421.32.Y45M67   1993
363.37'9—dc20   92-53358

93 94 95 96 97 ❖/RRD 10 9 8 7 6 5 4 3 2 1

To my family and friends—particularly my grandmother,
Sylvia Orloff, and Rob Hart
—and to firefighters everywhere

Man is the only animal that blushes. Or needs to.

—MARK TWAIN

Contents

Book Four: Paradise Lost

Maps

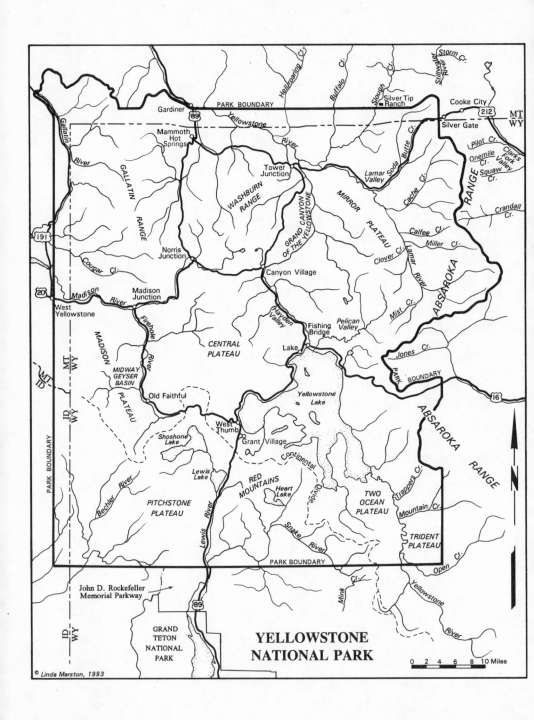

**YELLOWSTONE
NATIONAL PARK**

© Linda Marston, 1993

0 2 4 6 8 10 Miles

N

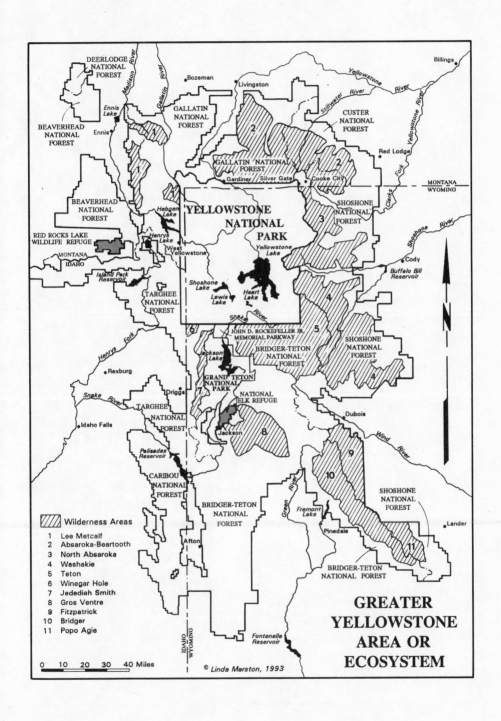

DEERLODGE
NATIONAL
FOREST

Billings

Bozeman

Livingston

Yellowstone   River

Ennis
Lake

BEAVERHEAD
NATIONAL
FOREST

Ennis

GALLATIN
NATIONAL
FOREST

2

CUSTER
NATIONAL
FOREST

Stillwater   River

Red Lodge

BEAVERHEAD
NATIONAL
FOREST

GALLATIN NATIONAL
FOREST

Gardiner / Silver Gate   Cooke City

2

SHOSHONE
NATIONAL
FOREST

MONTANA
WYOMING

Clarks   Fork

Hebgen
Lake

YELLOWSTONE
NATIONAL
PARK

3

RED ROCKS LAKE
WILDLIFE REFUGE

MONTANA
IDAHO

Henrys
Lake

West
Yellowstone

Yellowstone
Lake

Shoshone   River

Cody

Island Park
Reservoir

TARGHEE
NATIONAL
FOREST

Shoshone
Lake

Lewis
Lake

Heart
Lake

4

Buffalo Bill
Reservoir

Snake   River

5

SHOSHONE
NATIONAL
FOREST

Henrys   Fork

Rexburg

6

JOHN D. ROCKEFELLER JR.
MEMORIAL PARKWAY

BRIDGER-TETON
NATIONAL
FOREST

N

Snake   River

Driggs

Jackson
Lake

GRAND TETON
NATIONAL
PARK

4

TARGHEE
NATIONAL
FOREST

NATIONAL
ELK REFUGE

Idaho Falls

Dubois

Jackson

8

Wind   River

Palisades
Reservoir

9

CARIBOU
NATIONAL
FOREST

10

Green   River

SHOSHONE
NATIONAL
FOREST

BRIDGER-TETON
NATIONAL
FOREST

Fremont
Lake

Lander

Afton

Pinedale

11

Wilderness Areas

1   Lee Metcalf
2   Absaroka-Beartooth
3   North Absaroka
4   Washakie
5   Teton
6   Winegar Hole
7   Jedediah Smith
8   Gros Ventre
9   Fitzpatrick
10 Bridger
11 Popo Agie

BRIDGER-TETON
NATIONAL FOREST

GREATER
YELLOWSTONE
AREA OR
ECOSYSTEM

0   10   20   30   40 Miles

IDAHO
WYOMING

Fontenelle
Reservoir

© Linda Marston, 1993

Prologue

**Silver Gate
September 1988**

We were by the barricade at the road out of Silver Gate when I asked about the firestorm. Smoke and drifting ash obscured the heavily forested mountains on both sides of the town. Helicopters cut low overhead, trailing great buckets of water on steel cables. Earlier, the air tankers had been through, dropping retardant. Poor visibility would prevent their return.

I held out my hand and watched a tiny ember, a glowing pine-needle fragment, drift into my palm and turn to ash. The forest swayed before the approaching heat, dry and ready. Moving in from the distance was the sound of rolling thunder, the rumble of some cosmic freight train running hard down the vault of heaven. A million acres of the Yellowstone country had burned. The firefighters were retreating.

I watched them trudge past the barricade, their faces black with soot, dazed by sleeplessness, ill and beaten. They headed down the narrow forest road, away from Yellowstone Park, to nearby Cooke City. Flags flew from the small log cabins and lodges of Silver Gate, as if in parting salute to the firefighters. Within hours, Cooke City would be evacuated too.

We had been on the losing side for weeks, yet it was all still strangely compelling. There had been controversy and rage over the federal policies that had allowed fires to burn, charges of mismanagement, of delay and dark conspiracies "to burn down Yellowstone," but all of us recognized that we had come to be caught up in something much bigger than that now. History was being written in the flames. We had found ourselves in a fantastic war with nature in a lost country, a remote outpost of paradise on the faded frontier of the American West.

Through the acrid haze we caught sight of a high dark column of smoke leaning across the sky. That was when I asked about the firestorm. A Forest Service official explained that a firestorm is a kind of self-sustaining mobile world of destruction. It is a fire on the move, so large—miles wide and deep—that it creates its own wind and weather. Its mushrooming smoke column rises to 30,000 feet; giant flames flash up through the column as unburned gases ignite; ice crystals form in its upper reaches.

A firestorm is always searching for fuel. Before it hits, there is a lull, an eerie pause, and the trees gently sway and ash lightly falls, and then the ash comes faster and thicker and soon it seems like a snowstorm except that it is not snow and then there is that sound of a giant freight train bearing down on you and the wind picks up and the swirling ash turns to fire. The storm rains fire on the forest. Embers the size of fists fly out of the advancing front and the wind nears gale force now, hotter and harder than anything you've ever known, searing your lungs. Daylight is smothered by the smoke and a black sunless roaring surrounds you and fire blasts out of the darkness and if you are in the wrong place or if you panic you will certainly die because here comes a 300-foot-high wave of flame crashing through the forest, incinerating the underbrush, exploding in the treetops and destroying everything in its path.

It was, in short, one unholy son-of-a-bitch, and it was headed our way.

BOOK ONE

Yellowstone Nation

The Crow country is a good country. The Great Spirit has put it exactly in the right place.

—ROTTEN TAIL, A CHIEF OF THE CROW NATION, c. 1830

I

Clover

Dan Sholly, defense minister of the Yellowstone Nation, remembers racing toward the Lamar Valley in the SA-315 Llama helicopter. The thick forests of Yellowstone's Central Plateau fell away behind him as his pilot steered for the high peak of Mount Washburn. Sholly was on the move, out of the office, headed into action, and bliss it was that day to be alive! Sholly had it all! Officially, the Marine veteran was chief ranger of America's first and most famous national park, but the official title did not come close to describing his true role: he was the sheriff of Wonderland, the top cop of paradise, and with his radio network and armed rangers and cars and horses and snowmobiles and aircraft he commanded a ready force that warred constantly against the powers of evil, ignorance, and plain old cowboy mischief.

In an unheroic age, the rangers still carried a mystique of heroism and adventure. They searched for and rescued tourists who had fallen off cliffs, into rivers or hot springs, had been gored by buffalo, mauled by bears, nipped by coy-

4   otes, hit by cars, poisoned by water hemlock, or simply had wandered off lost and cold and miserable in the backcountry. Most of them they brought out alive. They arrested drunks and dopers and thieves. They hunted poachers. They dispatched troublesome bears—with a bullet when necessary, more often with a tranquilizer dart and a quick helicopter trip to a remote locale. Sholly made sure the roads ran free and the trails were sound, that nobody pissed in the streams or built campfires in "undesignated areas" or otherwise trammeled "the resource"—the park, that is, his universe, the Yellowstone Nation, 2.2 million acres of storybook meadows and glens, sizzling hot springs and geysers, secluded valleys, hidden lakes, trout-rich rivers, mountain refuges of lion and elk and bear.

No wonder Sholly quit that desk job in Washington as chief of *all* rangers! This was where the action was and Dan Sholly was a man of action. He protected the park from the people. He protected the people from the park. And when the policy allowed it, Dan Sholly fought fires.

Sholly recalled not being "overly concerned" about forest fires that day as he sped toward the Lamar Valley, Yellowstone's grand northern range. Sholly loved those helicopters. They were action incarnate. "Sholly's trolley," insiders called the park's specially leased helicopter, out of the chief ranger's hearing, or, less kindly, "Sholly's folly." The man himself they called—also out of earshot—"Danbo," after Rambo, and said he went to sleep wrapped in the American flag, which was an exaggeration, and whispered that the glass eye he had earned in Vietnam was etched on the back with the Marine Corps insignia, which was not an exaggeration.

He was a compact man, tough, with a romantic attachment to the can-do individualism of the Old West tempered by years in the bureaucracy and the scars of Vietnam. "A real warrior type," said his boss, Park Superintendent Robert Barbee. "The kind you would want on your side in a fight." Sholly was respected by most inside the park, and disliked and despised by many outside the park—a key distinction, that, inside the Yellowstone Nation and outside it. Those who despised him tended to be poachers and outlaws, while those who merely disliked him were reacting more to the symbol of Yellowstone than the man, for he was not an unkind person.

Like Hemingway's Robert Jordan, the doomed hero of *For Whom the Bell Tolls,* Sholly was schooled in the arcane arts of the wilderness. In fact, Hemingway had given the fictional Robert Jordan a home in the Yellowstone area. Hemingway's mystique imbued many in the high country—the rangers and the firefighters, the cowboys, the hunters, the backcountry enthusiasts, the loners seeking refuge from urban America. Of course, this had little to do with Ernest Hemingway the writer. It spoke instead to a shaping mythos of which Hemingway was keenly appreciative—the mystique of the capable man, the wily outdoorsman, an American archetype as old as Natty Bumppo and Huck Finn. Wildland firefighters read a lot of Hemingway. For them, there is still a frontier country, a place to test oneself, to pit oneself against the high and wild, to work with grace under pressure. The great fires of 1988 would provide many such tests.

As one might expect, Dan Sholly liked to keep his hands on all the levers of power. This would prove increasingly difficult as the summer of fire slid toward chaos. At that particular moment, however, on July 14, 1988, at 2 P.M., as Sholly flew into the Lamar Valley the vast machinery of the Yellowstone Nation was humming along at top speed.

It was a typically busy summer, the chief ranger recalled. He was supervising the "normal" flow of tens of thousands of tourists through the park; chairing Yellowstone's Fire Committee and monitoring the progress of eight small, lightning-caused "let-burn" fires in the park; and preparing for the "secret" vacation visit of Vice-President George Bush, newly nominated candidate of the Republican party for the presidency of the United States. Aloft over Yellowstone, Sholly monitored through his helmet speaker the aviation frequency, the park administrative frequency, the fire frequencies, and the helicopter intercom. Everything seemed to be under control. But the high and wild was about to give Dan Sholly a nudge, a little reminder about who, or what, was really in con-

6   trol. One of those small "let-burn" fires was about to blow up real bad.

"Let burn" would come to be a hated term in the precincts of the Yellowstone Nation, symbolizing the ignorance of the media and the masses in not appreciating the natural role of fire in "regulating" the Yellowstone "ecosystem." "Let burn" was the widely adopted shorthand for the Park Service's natural-fire component—technically known as "prescribed natural fire"—of a broad and increasingly controversial philosophy called "natural regulation."

In *Guardians of Yellowstone*, a memoir published after the fires, Sholly explained the natural-regulation philosophy: "Yellowstone was not meant to be a regulated collection of animals or plants like those in a conservatory or ranch. Instead it was supposed to be more of a preserve, where nature's players could interact undisturbed."

Fire is one of nature's major players. In 1972, a century of fire-suppression activities inside Yellowstone Park ended and was replaced by a policy that allowed most naturally caused— that is, lightning caused—fires to burn. "Letting nature again take over and burn the forests, as it had been doing since the last ice age," Sholly wrote, "was simply another logical step in the ongoing attempts of the Park Service to return the park's ecology, as much as possible, to its original state."

Earlier that day, on his way to Cody, Wyoming, for a logistics meeting with the Secret Service team handling the vice-president's visit, Sholly had circled the small Clover and Mist fires in a fixed-wing aircraft. Bush and campaign chairman James Baker had been planning to camp where the two fires were now burning. Sholly also made a quick flyover of the other small fires in the park. The experienced firefighter saw no unusual fire behavior. Everything was in the limits of the park's fire-management plan: the fires were creeping along the ground in areas no larger than a hundred acres, they were remote, and they posed no immediate threat to people or buildings. Yellowstone would let them burn.

A lightning strike on a hillside of lodgepole pine and

spruce in the backcountry near Clover Creek had started the Clover fire three days earlier, on July 11. Two days prior to that and 10 miles south, a strike on a steep slope of lodgepole and whitebark pine above Mist Creek had ignited the Mist fire—"the poetry of the park's geography," in the words of *Denver Post* columnist Jim Carrier, being used to christen the fires.

The Mist fire was so small and remote that it was only monitored by a lookout station for a week, but the Clover fire was close to trails and campsites. The park immediately prepared a Fire Situation Analysis (FSA) on the Clover fire. The report was ready twenty-four hours after the fire began.

An FSA is the first flutter in the blizzard of paperwork that attends a major fire. It is a quintessentially bureaucratic document, achieving the requisite dullness even while its subject is thrilling and dangerous. Contemporary firefighting is a weird amalgam of action and bureaucracy, and the FSA provides one of the keys to understanding it. In the case of the Clover fire, the FSA foreshadowed many of the problems that were to plague Yellowstone through the summer. Essentially, the FSA is a short document that estimates a fire's size, its potential for growth, and outlines fire-management options. It provides information on current and expected weather patterns. It analyzes the "fuel types"—the kind of trees and their age, the types of underbrush, meadows, bogs, and other features—in the fire area. It notes the "constraints on suppression activities" existing under administrative and legal guidelines, and states the basic thrust of the fire-management policy in force in the zone. "Social or external factors" beyond the immediate fire-management zone are considered, and a brief analysis of the environmental, economic, and social effects of both the fire and the possible suppression efforts is included. Alternative firefighting plans are sketched, and the costs of these alternatives are noted—Washington, and in the end the taxpayer, pays the bill for firefighting, and budget battles are fierce within the many government agencies responsible for firefighting, principally the Department of Agriculture and the Department of the Interior.

Appended to the FSA is a map of the fire area, lists of suppression forces, and costs under the firefighting alternatives. Finally, a decision—to be approved by the superintendent in the case of a national park fire and by the forest supervisor in the case of a national forest fire—is reached on what course of action to take. The FSA is a comforting collection of papers, full of facts and plans and alternatives, brimming with technological competence. The comfort begins to fade a little when one confronts the document that often follows the Fire Situation Analysis. It's called an *Escaped* Fire Situation Analysis.

Yellowstone Park's July 12 FSA pegged the Clover fire at 150 acres. The weather was hot and dry, humidity was low, and gusty winds with evening thunderstorms were anticipated. The thunderstorms were expected to carry little rain. (This weather pattern would be repeated through the summer.) The fuel types were "mature" lodgepole pine, spruce, and fir—typical of the Yellowstone area and ripe for burning—running in a continuous stand for 10 miles to the treeline below the eastern boundary of the park, at the alpine ridge of the Absaroka Range.

On the other side of the ridge was the Shoshone National Forest, the Silver Gate and Cooke City area, and the ranches of the Clarks Fork Valley and Sunlight Basin. Under "Important Social or External Considerations," the Clover FSA noted the possibility of "smoke and fire" into these areas, although no one in the park in July believed the fire would come anywhere near the Clarks Fork Valley or Cooke City. Rain would stop the fire, or the mountains would. What the local residents, outside the borders of the Yellowstone Nation, believed was a different matter, as we'll see.

The FSA estimated that the Clover fire would grow to a maximum size of no more than 25,000 acres, all of it inside the national park, "if no significant rainfall occurs." The FSA then outlined three "fire-suppression alternatives." Alternative A called for allowing the fire to burn, with careful monitoring on the ground and by air. The cost of the operation was estimated at $52,000, with no rehabilitation costs for seeding and replanting areas damaged by humans and machinery. Alterative A was given a 60 percent chance of success.

Alternative B was to fight the fire with two twenty-person crews and a helicopter for a week, followed by one crew for two weeks of mop-up, patrol, and rehabilitation. Suppression costs were estimated at $367,000; estimated rehabilitation costs: $10,000. Alternative B was given an 80 percent probability of success. Alternative C was to hit the fire hard with one of the nation's top firefighting units, known as a Type I team. This plan called for thirty crews for five days, a massive base camp, a forward firefighting post called a "spike camp," two light helicopters for two weeks, two medium helicopters for one week, and two follow-up crews for patrol and mop-up. Suppression costs: $2 million; rehabilitation: $50,000; probability of success: 90 percent.

Yellowstone's Fire Committee—Superintendent Robert Barbee, Chief Ranger Dan Sholly, Fire Behavior Analyst Don Despain, and Fire Management Officer Terry Danforth—chose Alternative A. In other words, they chose natural regulation. They would let it burn. "Fire is a natural process," the July 12 FSA concluded, "and the Clover fire should be allowed to burn naturally within the park. Suppression efforts"—bulldozers, tractors, trucks, control lines of cleared timber, base camps, spike camps—"can be more damaging to the environment in the long run." A "light hand on the land" approach was recommended, including monitoring the fire and protecting a few remote cabins used by patrolling rangers.

STANDARD FIREFIGHTING ORDERS:

Fight fire aggressively, but provide for safety first.
Initiate all action based on current and expected fire behavior.
Recognize current weather conditions and obtain forecasts.
Ensure instructions are given and understood.

10   **O**btain current information on fire status.
          Remain in communication with crew members, your
               supervisor, and adjoining forces.
          Determine safety zones and escape routes.
          Establish lookouts in potentially hazardous situations.
          Retain control at all times.
          Stay alert, keep calm, think clearly, act decisively.

—*FIRELINE HANDBOOK*, NATIONAL WILDFIRE
COORDINATING GROUP

In Cody, the Secret Service advance team had not been happy. Protecting the vice-president in unfamiliar wilderness was difficult enough without the additional problem of a burning forest. Sholly reassured them that the fires were not an immediate threat and that security for Bush and Baker was in place. He then flew back to Mammoth, switched to his helicopter, picked up "helitack"* foreman Dick Bahr, and lifted off for a fire-strategy meeting at Grant Village, the big visitor development on the west shore of Yellowstone Lake.

They never did make it to Grant that day. On the way, the Mammoth communications center relayed up a report from a backcountry ranger: Clover was on the rampage. In the few hours since Sholly had flown the fire, the winds had kicked up and the 300-acre blaze had expanded to 4,700 acres. Sholly was stunned by the ferocity. Thousands of trees, he later wrote, literally were "exploding," turning into "swirling torches of fire." A 20,000-foot smoke column—the first of many—was rising over the Lamar Valley. This was fire behavior more typical of late August or September in a dry season, not the moist high country in early July.

As Sholly, Bahr, and helicopter pilot Curt Wainwright would later recount in separate interviews, there were two immediate concerns. An outfitter, guiding some customers on a horsepack trip, was at nearby Miller Creek, right in the path of the fire. There was time to warn them to move on. But there might not be time to save the Calfee Creek patrol

* A glossary appears at the back of the book.

cabin, a historic structure used by backcountry rangers for generations.

The helicopter spun over the fire. Curt Wainwright remembers that through the smoke they could see a wall of flame consuming the forest, moving down a steep wooded slope on the west side of the Lamar River. The cabin was on the east side of the river, less than a mile away. From the main fire front emerged long fingers of flame, scuttling toward the cabin. They would have to act fast.

By chance, three of Dick Bahr's firefighters, members of an airborne attack unit called a "helitack crew," were working 3 miles away. They had been sent out that morning to shore up a dangerous patch of trail in preparation for Bush's vacation. This was not the usual business of a helitack crew, but VIP visits called for some rule-bending—after all, it just wouldn't do to have a horse carrying the vice-president of the United States slip from the steep trail and go tumbling into the river.

Sholly radioed the firefighters and told them to stand by in a clearing where they would be picked up to help protect the Calfee cabin. Wainwright touched down at Miller Creek and Sholly told the outfitter to move his party to a safe site 8 miles away. Then they lifted off and picked up the helitack crew.

Dispatched for trail work, not firefighting, the helitack crew was not fully prepared for the task ahead. They had not taken their initial-attack firefighting packs with them that day. In an early sign of a recurring problem, the helitack crew was experiencing an equipment shortage. Two crew members, Jane Lopez and Kristen Cowan, were not carrying an important weapon in the firefighter's arsenal: the lightweight, super heat-resistant "fire shelter."

The lifesaving device looks like a floorless aluminum pup tent and is usually worn in a pouch looped to a belt. When a situation appears life-threatening, the firefighter is supposed to deploy the shelter, crawling inside and hunkering down in a preestablished "safety zone," such as a meadow, a burned-over area, or a bulldozed clearing while the fire passes above. The

12   fire shelter has saved many lives, but in the black humor of the fire community it is known as a "shake and bake."

Regulations about carrying fire shelters were strict, and Sholly remembers having "an inkling" that he might be in for some trouble on this point. (He was correct.) But it was too late: the equipment he had requested from Mammoth would not reach them in time to save the Calfee cabin. He decided to risk it. They flew back into the fire and Sholly sized up the situation.

The Clover fire was sweeping down to the river. Yellowstone's prevailing southwesterly winds seemed to put the main firefront on a course that would pass slightly south of the cabin if the fire jumped the river, although a tricky down-canyon wind, moving from south to north, down the Lamar River, could carry the blaze to the cabin.

They dropped Bahr with a radio on a ridge a mile northwest of the cabin to serve as a lookout and flew back into the fire zone. As they attempted to land, Sholly wrote in *Guardians of Yellowstone*, the helicopter began to violently rock and pitch and there was the "sickening howl" of a windstorm around the craft. Sholly glanced at the others. "Their eyes were staring wildly at the thick grayish smoke and large embers swirling madly by. I tried not to think of the stalled rescue helicopter that years ago had brushed past my face and exploded into flames" at the bottom of Yosemite National Park's El Capitan Peak.

"I pulled the craft up, out of the storm," Wainwright later recalled. "The winds were too wild, too erratic. The craft was too heavy. We would have to reduce our load or we'd crash while attempting to land."

Sholly decided to let Jane Lopez off in a meadow 5 miles away. They touched down and Lopez scrambled out. The helicopter immediately broke for the sky—time was running out—leaving the firefighter with a radio but without provisions, her pack still lashed to the basket of the helicopter. "She looked so alone and vulnerable," Sholly would later write, "as if we had dropped her a million miles from civilization." Wainwright promised to come back for her if the fire neared.

Thick smoke had filled the valley around the cabin. Wainwright steered into it, bringing the ship down on a gravel bar beside the river. Sholly, Cowan, and helitack crew member John Dunfee leaped from the helicopter, gravel and grit and smoke and fire churning around them. "For a second I was back in Nam," Sholly wrote.

"I told Sholly I couldn't stay long," Wainwright recalled. "If the fire got much closer I'd have to pull out and fly down the river."

They ran for the old log cabin. It was tucked in under a stand of lodgepole pine and fir, with split firewood stacked against it: a bonfire waiting to go up. The crew would need to scatter the firewood, chop away overhanging branches, burn the closest trees to deny fuel to the advancing fire, and scratch a dirt line around the cabin to retard ground flames. "What I was doing," Sholly later said, "was attempting to fireproof the cabin." He took the radio from his belt and called in a request for an air tanker to make a retardant drop on the structure.

First, a safety zone needed to be established. Sholly led the crew to the meadow where they had initially tried to land. If the firestorm hit, they would flee to the meadow and get under the fire shelters.

With the wind howling around them, they dashed back to the cabin. *Move move move!* Firewood flew away from the walls. Sticks and pine needles were piled around the near trees and Sholly ignited a "fusee"—a hand-held flare used for starting fires—and turned it to the base of one of the trees. For a second, nothing happened, then flames crackled up the bark and into the branches. As often happens in wildland blazes, Sholly was using fire to fight fire.

Meanwhile, Curt Wainwright was waiting on the gravel bar, the sound of exploding trees growing closer every second. "Remember," said Wainwright, "I couldn't see much because of the smoke. But I heard Bahr on the radio talking to Sholly. The fire was crossing the river. The trees began to shake around me. I cut in on the frequency and told Sholly I'd have to leave in sixty seconds. Did he want me to pull them out?"

"Negative," Sholly replied. He was in his element. His

14   mission was to protect the cabin. "If we had pulled out," Sholly said later, "smoke would have prevented our return and the cabin would have been lost. I told Wainwright to get out while he could."

Wainwright lifted off, picked up Bahr, and circled the area. The fire was about to hit the cabin. According to the official investigation of the incident, Bahr radioed Sholly and again offered to pull them out. Again the reply came back: negative. Wainwright scooped up Lopez and for the third time offered to come in for the crew. Again, Sholly declined.

Suddenly a ball of fire exploded in a pine tree near the cabin. Large embers were "spotting" out ahead of the firefront. Flames from the pine leaped from tree to tree, turning them into columns of fire. They ran for the safety zone. To the west, the main fire was almost upon them; to the east, the fire had spotted into the trees. When the full force of the main wall hit, the powerful winds likely would suck the eastern burn toward the main front, catalyzing a bridge of fire over the meadow.

It was time to deploy the shelters. John Dunfee crawled into his, pulling the edges down tight to keep out the smoke and superheated air. Sholly spread out the other shelter and motioned Cowan in. The young woman was doing fairly well, Sholly recalled, considering the circumstances. She was not an experienced firefighter. In fact, she had only recently completed her training; this was her first fire. Sholly followed her into the shelter.

Sholly and Cowan lay on their stomachs, holding down the sides of the shelter with their boot tips and gloved hands. The wind whipped at the thin walls, pelting it with ash and embers. The heat was sucking away the oxygen. Sholly saw that the woman was, understandably, nervous.

"I sought to keep Cowan calm by talking to her," Sholly noted in his memoir. "Always in the background, however, were the mocking voices of the storm: the terrible roar of tree crowns being snatched by the fire, the sharp crack of limbs being ripped off and slammed to the earth by the fire's own gale-force winds, and the distant reports of boulders shattering into innumerable shards under the blast-furnace heat."

Clover raced across the treetops—"crowning out," in firefighter's parlance—waves of flame curling forward to absorb tree after tree, pausing for a moment as the newly released energy of the captive branches pushed the flames upward, then surging forward again. The sound of cracking trunks and root systems heaving out of the earth punctuated the constant deep rumble. Then came a big explosion: to Sholly it sounded like the ranger cabin going up.

After forty minutes, the great winds began to subside. They had made it. The fire front was rolling on. Sholly poked his head out from under the shelter. The smoke and overpowering smell of burning wood bit at his eye and lungs. The green meadow and forest were gone. The earth was scorched, a ghostly pall of smoke hung in the air, and all around them were the skeletons of blackened trees and flickering spot fires. Incredibly, the cabin had been saved. Their fireproofing had worked.

They stayed on the site a few more hours, cleaning up around the cabin and putting out the spot fires. Sholly would later write that what awed him was not so much the *size* of the firestorm but the *way* it had burned. "It had attacked everything in its path with a ferocity that is more common in places as dry and brittle as southern California." By the time Wainwright returned and they lifted off from the meadow, its perimeter eerily glowing in the night, Sholly had made a decision.

He flew back to Mammoth and called together the Fire Committee. Superintendent Barbee would be presented with their recommendations after the meeting. They met at 9:30 P.M. in the fire cache, Yellowstone's emergency-operations center, a low stone structure behind the administration building. Of the eight "prescribed," lightning-caused natural fires inside the park, Clover was the biggest, at roughly 4,700 acres. Also in the eastern sector were two other fires: the Mist fire, burning at 200 acres just south of Clover; and the Raven fire, like Mist

16   a July 11 lightning-start, isolated at one acre southwest of Mist. The big fire in the northwest sector was the Fan fire, burning by July 14 at about 2,900 acres. Seven miles southeast of Mammoth, on the Blacktail Deer Plateau, the Lava fire was an acre wide.

Three natural fires were burning in the south. The Shoshone fire, at 70 acres, was pointed in the direction of the Grant Village development 9 miles away. South of the Shoshone, the Red fire had burned to Lewis Lake and was roughly 680 acres in size. Down by the southern border of the park, the Falls fire, two days old and 25 acres large, was making the neighboring Targhee National Forest nervous. The Targhee did not want the Falls fire on its property. (Careless campers trying to burn toilet paper had started a ninth fire, the Narrows, north of Shoshone Lake, earlier in the day. The 2-acre blaze was not a naturally caused fire and under park policy it would be fought right away.)

Don Despain, a park biologist and one of the fire-behavior specialists on the Fire Committee, believed that all the fires, except Narrows, still should be allowed to burn. This was what natural regulation was all about.

But it wasn't just about natural fire ecology anymore. The public clamor was beginning to grow, and the fires were starting to become entangled in regional politics. The Falls fire was threatening to cross into the Targhee National Forest in the south. The Clover fire was approaching Yellowstone's eastern border with the Shoshone National Forest. In the north, the Fan fire was moving toward the Gallatin National Forest. All three forest supervisors—Barbee's counterparts in the U.S. Forest Service—were not inclined to "accept" the fires onto their land. Firefighters would soon be enmeshed in complex maneuvering between Park Service and Forest Service officials.

Beyond the borders of the Yellowstone Nation, other fires had started too. South of Yellowstone and next door to the Targhee National Forest, lightning had started the Mink fire on July 11 in the Bridger-Teton National Forest. At first the Mink fire had been allowed to burn, but commercial interests and congressmen were pressing for action; firefighters would soon be on the way. Northeast of Yellowstone, firefighters were attacking a blaze in the Custer National Forest.

Nature had begun to tie the first knots in a noose of fire. Simultaneously, internal and external forces were driving Yellowstone to action, forcing it to move beyond the monitoring and planning stages of the fire-management plan. Inside the park, Sholly and some other officials were alarmed by the fire behavior and worried by the lack of rain. Outside the park—in tiny "gateway communities" like Silver Gate and Cooke City, and in nearby cities such as Cody, Bozeman, and Billings—a perception was taking hold: While the Bridger-Teton and Custer National Forest fires were being fought, the fires inside Yellowstone National Park were burning unchecked. At this point in the summer, it was an accurate observation.

On the night of July 14, Sholly argued for and won the Fire Committee's approval to begin to take steps to fight some of the fires. The chief ranger still believed in natural fire, but things were starting to get out of control. They would take measures to start fighting the two southern fires that threatened Grant Village and the Targhee National Forest, the Shoshone and Falls fires. The Clover fire, however, the fire exhibiting extreme behavior, the fire roughly 18 miles from Silver Gate, Cooke City, and the homes of the Clarks Fork Valley—a hard 18 miles, up thickly wooded creek drainages and over the 9,000-foot ridge of the Absaroka Range—was allowed to burn.

But on July 14, allowing Clover to burn was not seen as a fateful decision. The fateful move, in the eyes of the Fire Committee, was their agreement to recommend to the superintendent that Yellowstone move to a suppression effort, using an Incident Management Team, under the Incident Command System. Essentially, Yellowstone Park was about to call up an army. Drawing from the national ranks of professional wildland firefighters, the Incident Command System brings together a general—the incident commander—with a trained staff and troops. The staff, including chiefs and deputies of operations, planning, logistics, and finance, is the brain center

18   of the firefighting effort. It's called a Type I—or in less critical cases a Type II—Incident Management Team, or "Overhead Team." The troops of the Type I or Type II teams include grunts—infantry-style "hand crews"—and might include crews manning pumper trucks called "engines," bulldozer crews, sophisticated aviation units with smoke jumpers and helicopters and bombers, medics, communications specialists, scouts, cooks, timekeepers, and other backup personnel, as the situation demands. The fire itself is the "incident."

Initially, Dan Sholly would be the incident commander for all the Yellowstone fires and his troops would be drawn from Yellowstone and other National Park Service personnel. Moving to the Incident Command System was an acknowledgment that it was no longer a normal summer; Yellowstone had yet, however, to acknowledge that it would need outside help. Sholly had fought fire for twenty-five years and was a qualified Type II incident commander. But soon the fires would become too big, too complex, and overall control would slip from his grasp. Other generals from the national network would come in and the command structures would multiply—and so would the confusion.

But all that was in the unseen future. Following the Fire Committee meeting, Sholly walked across the quiet compound to report to the superintendent. Barbee somberly listened to him and approved the Fire Committee's recommendations. Barbee recalls it as "a dispiriting day." Earlier, he had telephoned the supervisor of the Shoshone National Forest and advised him to cancel George Bush's trip from the Shoshone Forest into Yellowstone Park. The future president thus became the first visitor turned away from Yellowstone because of the fires. Later that day, it had seemed to Barbee that "my chief ranger was about to be fried alive in Calfee Meadow." Now, following Sholly's report, Barbee saw that the park was starting a gradual retreat from the natural fire component of the natural regulation policies. "I had been defending natural fire since late June, from the ignitions of the fan and Shoshone fires," Barbee recalled. "But now I saw that we were beginning to get into some real problems."

What Barbee and Sholly did not realize was that their troubles were only just beginning. The public would remain confused about the park's firefighting intentions for weeks to come. The national press corps would descend in a feeding frenzy. And the blazes would rip through the summer, rewriting a generation's expectations of fire behavior. The stage had been set for one of the greatest wildland fire battles in American history.

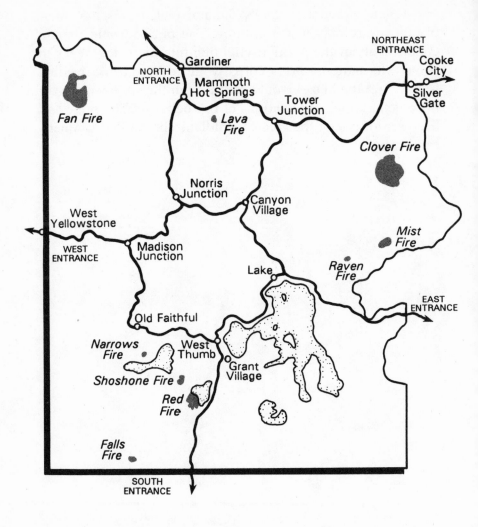

YELLOWSTONE FIRES

July 15

Park issues its first fire map.

II

Mammoth Hot Springs

The Yellowstone fires were a drama of power and politics—at once Greek tragedy and comedy of errors—acted out against an Old West tableau of epic scale. At the center of the political controversy was Superintendent Robert Barbee and the "let-burn" component of Yellowstone's natural regulation policies. Many forces competed for power, but in the end it was fire itself that ruled the season.

Indeed, long before the white man arrived at Yellowstone, fire was an integral part of the region. Little is known about the activities of prehistoric man in the area except that he was a hunter, moved in small bands, and used fire. Man has been traveling—and burning—the Yellowstone country for at least ten-thousand years.

Around Yellowstone the main descendants of the Paleo-Indians were the Crow, Blackfoot, and Shoshone tribes. More is known about their habits, and not all of it is pretty. Some of their uses of fire might shock those who celebrate the Native American as an ecological angel in mystical harmony with nature.

Fire, after all, is great fun to watch—there's a touch of the

22   pyromaniac in every firefighter—and the Indians were not above torching off a few trees to provide some spectacular evening entertainment, a scene recorded by Lewis and Clark. Not having had the benefit of an education by Smoky Bear, Native American hunting and warring parties usually left campfires burning when they moved on. Fire drives were used in hunting: fields would be set ablaze and the terrified buffalo driven off cliffs. Much of the meat was left behind to rot.

Although the Great Spirit of environmentalism cannot be said always to have been with the Native American, in many ways he used fire wisely—and taught the early frontiersmen to do the same. The historian Stephen Pyne writes in his encylopedic study, *Fire in America,* that it "is hard now to recapture the degree to which Indian economies were dependent on fire. In its domesticated forms, fire was used for cooking, light, and heat. It made possible ceramics and metallurgy. Its smoke was used for communication. It felled trees and shaped canoes." It was used to improve pastures for horse and buffalo, to "fireproof" fields around villages and open trails through thick underbrush.

As soon as the white man began to push West, he adopted Indian fire practices. Around 1830 the first white man showed up in Mammoth Hot Springs with a band of Crow. These early trappers and explorers usually had a wary respect for the Indian and followed Indian fire practices in flushing game and fireproofing campsites. Many of the early travelers report encountering vast unchecked blazes or evidence of recent fires.

By the mid-1840s, the famed explorer Jim Bridger was spreading tales of a distant land of underground fire, full of "spouts" and "great springs, so hot that meat is readily cooked in them." That was a fairly reliable bulletin, but it was from the same man who told of a mountain that had been cursed by a great Crow medicine man—every living thing on it had been turned to stone. Aubrey Haines recounts Bridger's tale of the discovery of the petrified forest of Yellowstone Park's Specimen Ridge in his two-volume history, *The Yellowstone Story.* "Come with me to Yellowstone next summer," Bridger told a friend, "and I'll show you petrified trees a-growin', with petrified birds on 'em, singin' petrified songs." Embroidery such as that led Bridger's tales of the marvels of the Yellowstone high country to be dismissed as the hallucinations of a rogue too long away from civilization.

But the rumors of a fantastic land in a mountain stronghold persisted, and soon a wave of organized exploration would open the Yellowstone high country. With that opening would come the first salvos in a historic argument that persists to this day, and found a strange new mutation in the great fires of 1988: whether to preserve the Yellowstone area's riches or to keep that wealth—the timber, game, grasslands, gold, silver, gas, oil—open to those with the strength and endurance to go in and get it. A trail of fire traces through this historic quarrel. Let us follow it for a moment.

In the summer of 1870 an expedition of Gentlemen led by Civil War veteran Henry D. Washburn, surveyor-general of the Montana Territory, confirmed that the wonders of Yellowstone were not merely fictions advanced by rascals like Jim Bridger. A day after stumbling onto Old Faithful, the Washburn Expedition made camp beside the Madison River. It was there that "a rather unusual discussion" took place, noted the Montana politician Nathaniel P. Langford in his diary, later published under the title *The Discovery of Yellowstone Park*. Some members of the expedition wanted to claim the area and divide it up between themselves. But the journalist Cornelius Hedges, reports Langford, suggested "that there ought to be no private ownership of any portion of the region, but that the whole of it ought to be set apart as a great National Park."

It was an idea whose time had come. Thanks to the publicity given to the Washburn Expedition and the lobbying efforts of its members, Congress in 1872 passed the Yellowstone Park Act. The area would be "set apart as a public park or pleasuring-ground for the benefit and enjoyment of the people." Langford was appointed its first superintendent. Over time, treaties were broken, and the Crow Nation, along with other smaller tribes, were pushed from the lands in and

24   around the new park, lands the Great Spirit had put in exactly the right place.

A revolutionary new nation—a Yellowstone Nation—was being born. The revolution had unfairly displaced the Indians, but it had given birth to the idea of a national park. For although, as Aubrey Haines has noted, the notion of a "park" was at least as old as the royal gardens of Persia (called *paradeisos*, paradises, by the Greek adventurer Xenophon) and the Norman *parcs* of feudal France, Yellowstone would be the first truly national park, the first park not reserved for the aristocracy—a park for the *demos*, the people. Like the idea of democracy itself, the national park idea would spread around the globe.

Yet almost immediately there was trouble in paradise. The *demos*, it turned out, had a lot of different ideas about what constituted their benefit and enjoyment. Congress had appropriated no funds for park administration and left unclear the law-enforcement role of the western territories in the new federal creation. Prospectors and others were trying to eke out a life at the fringe of the park, in places like Cooke City and in Gardiner, just north of the newly established park headquarters at Mammoth Hot Springs. They did not like the fact that park lands, under the congressional act, had been officially "withdrawn from settlement, occupancy, or sale."

For the next fourteen years, Robert Barbee's predecessors in the fledgling civilian administration skirmished with the citizens of the surrounding settlements, many of whom continued to cut wood, hunt, fish, and run livestock on the new preserve. Severely understaffed, park personnel also fought fires, blizzards, Indians, corrupt politicians, stagecoach robbers, scheming businessmen, drunks, brawlers, and a growing number of tourists, many of whom were hacking off chunks of geyser cones and carrying them away as souvenirs.

Expulsion was one of the few penalties that had any effect on transgressors. Naturally, this was extremely unpopular in the local communities, instilling a resentment that would linger long in the collective memory. On occasion, the exiles allegedly would take revenge by torching off a few acres near the superintendent's home at Mammoth.

That was the situation in August 1886, when the secretary of the interior, disgusted over congressional inaction while the park was being burned and vandalized, turned to the Army to rescue Yellowstone. (One hundred and two years later, as the final act of the great fires unfolded, the Army would return, proving that the more things change, the more they stay the same.) "I regret to have to report," Captain Moses Harris wrote the interior secretary after marching fifty men into Mammoth to take command of the federal preserve, "that destructive forest fires have been raging in the park during the greater portion of the present season."

The fires of the late 1880s began nearly a century of fire-suppression efforts at Yellowstone. As Stephen Pyne notes in *Fire in America*, Yellowstone firefighting had a broader significance too. The Wild West was being tamed. Yellowstone became a symbol of that pacification. Fire suppression, Pyne writes, "was a visible, material, and symbolic expression of Army determination to rid the park of destruction and vandalism of all sorts, to regulate tourism, and to confront and remove the lawless class of poachers. The regulation of people and the control of fire were complementary duties. It is ironic that the establishment of wilderness [with the federal act creating the park] meant the expulsion of frontier elements. But so it did."

The American frontier was vanishing, taking its restless explorers with it, but leaving behind their archetypal values. One of the first to recognize this was a young historian named Frederick Jackson Turner. With settlements dotting the continent, the country no longer had a line of frontier, over which lay wilderness, mystery, and danger. But, Turner realized, the country still had a frontier mentality.

In fact, it was America's westward movement and the "winning [of] a wilderness," Turner saw, that defined the American character. "To the frontier the American intellect owes its striking characteristics," he wrote in *The Significance of the Frontier in American History.* "That coarseness and strength combined with acuteness and inquisitiveness; that practical, inventive turn of mind, quick to find expedients; that masterful grasp of material things, lacking in the artistic but powerful to effect great ends; that restless, nervous energy; that dominant individualism, working for good and for evil, and withal that buoyancy and exuberance that comes with freedom—these are the traits of the frontier, or traits called out elsewhere because of the existence of the frontier."

One place these traits would continue to be called upon was in wildland firefighting. With the approach of the twentieth century, firefighters at Yellowstone and elsewhere struggled to suppress blazes with low manpower and poor resources. "Forest fires raged uncontrolled on every side of the park and destroyed millions of acres," reported Captain Frazer Augustus Boutelle, the second military commander of Yellowstone, in 1890. During an attempt to control a fire in the park's Gibbon Canyon with a small detachment of troops, for example, Boutelle noted that the climb up the fiery hills "was so difficult that two men had epileptic fits from the effort."

Resources were crude at best. "Up to a late date last season," Boutelle wrote his superiors in Washington, "there was no fire equipment in the park. The few axes and shovels supplied the troops for garrison purposes were the only tools available. Application was made for funds for the purchase of axes, shovels, and folding rubber buckets, but through some misunderstanding the authority was not promptly received, and the work was doubly hard from want of proper tools."

Boutelle was about to make a mistake that would sink many a Yellowstone Park superintendent. It's a lesson in the political culture of Yellowstone, one that Robert Barbee had absorbed long before 1988. Successful superintendents were politicians as well as leaders. Boutelle, the blunt frontier soldier, lacked the necessary skills. He made the mistake of reporting to Washington an encounter with a generous tourist at the height of the fires. Upon hearing that Boutelle "had applied for rubber buckets and had failed to get them," the visitor handed over $40 of his own, exclaiming " 'if this great United States Government or the Secretary of the Interior has not money to buy you a few rubber buckets for the protection of this wonderful and beautiful country, I have!' . . . Would that Congress," Boutelle concluded, "would take such an interest in the protection of the park before it is too late." Within a year, the impolitic Captain Boutelle was out of a job.

Yellowstone muddled on. It took a disaster to spur change in the world of wildland firefighting. Disaster came in 1910, when devastating fires swept across Montana and Idaho, burning over a million acres, wiping homes and towns from the map, and killing eighty-five people. The tales of hardship, courage, and endurance emerging from the 1910 fires gripped the American imagination. Out of 1910, Stephen Pyne has observed, came "a lore of high adventure amid a Lewis and Clark backdrop of wilderness splendor."

"I wish to preach not the doctrine of ignoble ease," wrote the rough-riding conservationist, Teddy Roosevelt, with the events of 1910 partly in mind, "but the doctrine of the strenuous life . . . to preach that highest form of success which comes, not to the man who desires mere easy peace, but to the man who does not shrink from danger, from hardship, or from bitter toil, and who out of these wins the splendid ultimate triumph."

While 1910, Roosevelt, and the romance of "the strenuous life" helped ignite a generation's interest in "conservationism"—a rough historical parallel to today's interest in "environmentalism"—the killer fires also brought practical and dramatic improvements to firefighting. State and federal fire-prevention efforts took on a new life. Congress passed the Weeks Act, enabling the U.S. Forest Service to enter into joint funding arrangements for fire protection with the states. "The effect," Pyne writes, "was to nationalize fire protection under the aegis of the Service." From then on, the U.S. Forest Service would be the top gun of wildland firefighting. Its view of fire would dominate the increasingly complex world of wild-

28   land management for the next fifty years: fire was bad; it destroyed valuable timber, homes, sometimes even human life; the goal was to suppress it as quickly as possible.

Following World War II, the Defense Department developed a keen interest in mass fire—a legacy of Dresden, Hiroshima, and Nagasaki. It began to work closely with the Forest Service. Joint studies on fire behavior and control were undertaken, research facilities were established, helicopters and air tankers and smoke jumpers came into use, and war surplus equipment was made available for firefighting. While manpower was still important, the Forest Service began to rely more on bulldozers, tractors, pumper trucks, and other heavy machinery. Armored and airborne firefighting divisions developed alongside the infantry-style ground troops. A militaristic if ragtag pursuit in its early days at Yellowstone, by the 1960s firefighting had become the province of a skilled warrior class.

Things, however, were about to change again, and it was these changes that would have the most profound impact on the great fires of 1988. The cultural shifts of the 1960s were finding their way into the policies and practices of wildland management and fire control—a new age, and a new attitude toward fire, was dawning.

The new attitude was part of the broad scientific and philosophical reevaluation that had given birth to contemporary environmentalism. On the scientific side, biologists were being drawn to the idea of self-regulating "ecosystems"; that is, the idea that a specific natural body, perhaps as small as a pond—or perhaps as big as the multi-million-acre land mass of Yellowstone Park and the areas surrounding it—might be seen as a largely self-contained, independent entity, an "ecosystem," capable of regulating itself if left alone by humans. Nature, in other words, could find its own "balance." Mankind's job in the national parks was to help nature regulate itself.

The ecosystem idea began its journey to center of the wildland policy with a 1963 report published by an Interior

Department commission chaired by A. Starker Leopold, son of the legendary conservationist, Aldo Leopold. The Leopold Report, as it came to be known, continues to guide Park Service policy. It recommended that "a reasonable illusion of primitive America could be re-created, using the utmost in skill, judgment and ecological sensitivity." The goal, in the Leopold Report's famous phrase, was to create "a vignette of Primitive America." Fire was part of Primitive America, and its role in the ecosystem was to be preserved.

Not coincidentally, at the same time the Leopold Report was being incorporated into National Park Service policy, philosophic ideas sharply at odds with Judeo-Christian culture were crystallizing in the environmental movement. For environmentalist intellectuals, the writer Lynn White, Jr., brought the problem into focus in an influential and widely reprinted 1966 essay, "The Historical Roots of Our Ecologic Crisis." White contended that Christianity, in contrast to ancient paganism, "not only established a dualism of man and nature but also insisted that it is God's will that man exploit nature for his proper ends." Man would subdue the earth, it was writ large in Genesis, and have dominion over every living thing upon it. The Judeo-Christian break from paganism, White argued, had alienated man from the animistic spirit— the belief that all natural phenomena have souls—and led to a degradation of the environment. "By destroying pagan animism," White wrote, "Christianity made it possible to exploit nature in a mood of indifference to the feelings of natural objects."

The snake in White's Garden was industrial civilization, capitalism, forward-rushing technology, forces that would bring about an environmental apocalypse if not restrained. "The Historical Roots of Our Ecologic Crisis" encapsulated all the paradoxes of the new wave of environmentalism—it was at once "progressive" and reactionary, embracing popular anticapitalist notions yet locating redemption in a preindustrial or antiindustrial world that couldn't possibly support global growth. White hadn't invented the new religion of environmentalism, but he wrote its first modern testament. Other

30   prophets would soon follow. The writer Roderick Nash would approvingly describe environmentalism as a new "gospel of ecology" propounded by "ecotheologians." The radical Earth First! environmental group leader David Foreman would preach that "human destruction of the wild" is the "keystone to understanding our alienation from Nature, which is the central problem of civilization." Here, then, was a philosophy, or theology, that seemed to fit snugly alongside the science of ecosystems and natural regulation.

At Yellowstone, the torchbearer of the natural regulation and natural fire policies was Don Despain. A research biologist specializing in plant ecology and an experienced fire-behavior analyst, Despain was no wild-eyed prophet. A gentle giant of a man and a player in the park's powerful science bureaucracy, Despain's mild demeanor concealed a passion for natural regulation in all its forms. He arrived at Yellowstone in 1971 and helped create one of the nation's first fire-management plans that included a significant role for naturally caused fires. The importance of this cannot be overlooked. The entire philosophic and practical foundation of wildland firefighting, in the world's most important piece of wildland, was shifting from total suppression to an emphasis on free-burning fire.

Written in 1972, revised in 1975 and 1987, Yellowstone's official Fire Management Plan placed virtually all of the park in a zone that allowed natural fire. (Areas close to major structural developments were excluded from the natural fire zone.) The plan was very much the child of the Leopold Report and the ecosystem idea. Its primary objective, as stated in the 1987 revision, was to "permit as many lightning-caused fires as possible to burn." The plan also called for protecting "human life, property, historic and cultural sites, specific natural features, [and] endangered species" from wildfire; and for suppressing wildfire—defined as fire that is freeburning or burning "out of prescription"—in "as safe, cost-effective, and environmentally sensitive a way as possible."

For the first sixteen years of the natural fire regimen, from 1972 through 1987, Yellowstone recorded 235 natural fires; even with the addition of some human-caused fires, only about 34,000 acres burned. Only thirteen of the natural fires were over 500 acres. There were two fairly active fire seasons: in 1979, over 10,000 acres burned; in 1981, over 20,000 acres. "We gained a lot of experience with large fires and numerous small fires during those years," Despain says.

For the most part, however, Yellowstone's pattern of cool wet summers persisted, although by the middle of the 1980s a drought was gripping many parts of the West. Yet in 1988 around the Yellowstone area, April and May rainfalls were above normal. By late June, things were drying up and Don Despain was "looking forward" to a season of fire.

It was not an emotion shared by everyone in the Yellowstone region. The attitude that fire was "evil" still had many supporters, particularly in the Forest Service. Six national forests surround Yellowstone Park; with one exception, the Beaverhead National Forest to the west, these forests directly border the park. Together, Yellowstone Park, the national forests, the gateway communities inside the national forests, and the small private land holdings form the "Greater Yellowstone Area" or, as some call it, the "Greater Yellowstone Ecosystem"—a term loaded with political as well as biological implications. The area, or ecosystem, measures anywhere from 6 million to 18 million acres, depending on who is doing the measuring.

Despain believed that the ecosystem idea was "good science" and found the "old-fashioned attitude" toward fire frustrating. Fire, to Despain, is part of the natural process. "Maybe it kills trees," the biologist explained, "but it clears the way for new trees. Maybe it burns up bird nests, but it creates conditions for new nests. Fire is part of the system and therefore it causes no damage to the system. In fact, it is fire suppression that causes damage by destroying the natural frequency of fire in the ecosystem."

Yet dissent from park management policies had not been limited to the old guard of the rival Forest Service and the historically antagonistic gateway communities. In 1987, a defector from the Yellowstone Nation surfaced with a bitter and

32   detailed study entitled *Playing God in Yellowstone: The Destruction of America's First National Park.* His name was Alston Chase and for years he had been the consummate Yellowstone insider: well educated, well connected, active in his support of the park, and a dedicated environmentalist. Chase had many friends in Yellowstone and by his own account he had set out to write a book about the success of the park's policies. But as his research proceeded, Chase became convinced that the policy of natural regulation was a nightmarish concoction of faulty science, bureaucratic cravenness, and environmental romanticism. His book stunned Yellowstone officials, who felt personally betrayed by an ally.

One of Chase's central and most persuasive arguments is that the ecosystem concept is more of a "quasi-mystical idea" than good science. In Chase's view, the first tentative scientific notions of a self-regulating natural body, advanced by botanist A. G. Tansley in 1935, were hijacked in the 1960s and 1970s by an environmental movement that preferred feelings over fact. Hard evidence that an ecosystem could regulate itself—or could even be defined—was lacking.

R. L. Lindeman's classic 1942 ecosystem-model study of the Cedar Lake Bog in Minnesota, "The Trophic-Dynamic Aspect of Ecology," had not survived the test of further scientific probing, Chase notes. Later scientists, Chase wrote, "began to see that the system was incomplete: even areas as apparently isolated as ponds were not closed systems. They were connected with the world in countless ways. Rather than isolated bodies of water, they were found to be merely part of the water table exposed above ground. The various forms of life that ponds contained were affected by underground currents, spring runoff, migrating waterfowl, trace elements dropped by rain, airborne spores, and the sun."

If something so apparently "simple" as a pond proved elusive of definitive ecosystem boundaries, what could be said of Yellowstone Park's 2.2 million acres and the vast landmass the proponents of wider preservation had dubbed the Greater Yellowstone Ecosystem? "The vaunted self-regulating ecosystem," Chase wrote, " . . . the supreme assumption that guided the management of wildlife in Yellowstone, more and more scientists began to suspect, was just a fictitious axiom, conceived to satisfy philosophical and mathematical conceptions of symmetry."

Yet while the ecosystem idea continued to be studied and redefined by scientists, its basic premise proved to be powerfully seductive in the popular culture. "It suggested to the lay public the fundamental insight of environmental awareness: that all things were interconnected," Chase noted in *Playing God*. "It gave substance to the growing sense—articulated by Thoreau, Muir, and other theologians of the environmental movement—that nature was a web of mysterious, hidden forces that the contemplation of the wilderness stirred in them. Consequently, the idea quickly became misunderstood. It became a name for these mysterious forces which bind natural objects together, and identified with that which evokes awareness of those forces: the apotheosis of unsullied nature, wilderness itself."

The seductiveness of the ecosystem philosophy was greatest among the counterculture activists of the 1960s and their political heirs, the men and women who would provide the fuel for the rapid growth of the environmental movement in the 1980s. The ecosystem philosophy became the religion of the environmental movement. The new missionaries would concentrate their activities not in the churches but in the arena of public opinion and public policy.

In 1983, Yellowstone found religion. Environmental activists from local chapters of the Audubon Society, Sierra Club, Defenders of Wildlife, Wilderness Society, Northern Rockies Action Group, National Wildlife Federation, and twenty-nine other groups formed what Chase called "the most ambitious effort by environmentalists to protect Yellowstone" in fifty years: the Greater Yellowstone Coalition (GYC). Many members of the GYC, particularly in its early days, were relatively new arrivals in the region; most were young, white-collar professionals, urban refugees; many were well-to-do and were looked upon with suspicion by longtime residents of the region.

34   The GYC was—and is— deeply hostile to Forest Service policies. While the Park Service operates under the preservationist ethic of noninterference with nature, the Forest Service is guided by a "multiple-use" ethic that allows many of the traditional economic pursuits of the region—such as logging, cattle grazing, mineral and oil extraction, as well as hunting and the use of recreational vehicles—to proceed in specified areas. As for fire, in most cases—with a few critical exceptions—in a national forest it is fought.

Like the administrators of Yellowstone Park, the GYC viewed natural regulation as the Alpha and Omega of wildland management. Nature could do no wrong. Yet the GYC often took matters a step further than the working professionals of the Park Service. Unlike the Park Service, the nonprofit interest group appeared to derive its central philosophical tenets from the "Deep Ecology" stream of the environmental movement, which viewed capitalist society as the root of evil in a world living out of harmony with nature. The answer was to give nature more room. The answer was more preservation, less multiple use.

The GYC played an important role in the political ecology of the Yellowstone area. Organized, articulate, aggressive, expansionist—it flanked the mostly middle-of-the-road liberals of the Park Service to the left. It could say and do what the Park Service could not. It lobbied. It sued. It directly fought the Forest Service. And, aside from a telling disappearance at the height of the blazes, it supported Yellowstone Park and the natural-fire program, providing the media with well-crafted sound bites.

As Alston Chase has noted, the GYC "had no bone to pick with the Park Service." Both envisioned a vastly expanded preserved region. The road to this environmental utopia, alas, would have to be strewn with the corpses of the traditional extractive industries. Logging was out. Mining was out. Oil drilling was out. Hunters and snowmobilers were a menace. The gateway communities were a blight on the land. The GYC located a kind of secular redemption in the "salvation" of the

Greater Yellowstone Ecosystem, an area that expanded in its official literature as the years went on.

"We're on the front lines every day," wrote GYC president Thomas McNamee, a wealthy New York author and occasional Montana resident, in the group's 1988 annual report, "beating back timber sales, heading off oil wells, fighting new roads, standing up for wildlife, analyzing new land use plans, testifying, pleading, negotiating, coordinating with our other member groups, drawing on their special expertise, dealing day after day, one by one, with the proliferating threats to Greater Yellowstone's natural integrity. . . . We are going to *save* the Greater Yellowstone Ecosystem." Wrote GYC activist George Wuerthner in Montana's *Billings Gazette:* "Save a forest; let it burn."

By the late spring of 1988, Robert Barbee was overseeing what was indeed a unique creation, if not quite the one its founders had envisioned. Yellowstone Park was the heart of a wildland political economy more powerful and more complex than anything Barbee's predecessors could have imagined, probably a wildland economy more complex than any on the globe.

Barbee ruled over an area of some 3,400 square miles, larger than Delaware and Rhode Island combined, with the largest concentration of wildlife in the lower forty-eight states and over ten thousand thermal features. Over 2.5 million visitors crossed his borders every year. At peak season, his combined Park Service and concessionaire work force numbered over 5,000. His combined budget, gate receipts, and franchise fees from concessionaires rounded out at roughly $20 million annually. The park drew at least another $20 million into the region in indirect income and jobs in the tourism industry. All this made Barbee and Yellowstone a force to be reckoned with in the precarious economy of the Northern Rockies.

Before the great fires, Bob Barbee and Yellowstone Park had few detractors. Alston Chase was one. Relations with both the gateway communities and the Forest Service came with some built-in tensions, but the problems were not seen as

36 insurmountable. The Greater Yellowstone Coalition was viewed as an important, if occasionally embarrassing, ally; preservationist ambitions for the Greater Yellowstone Area— reintroduction of the wolf to the ecosystem, the long fight against timber and mining interests, the addition of more protected wilderness areas—were slowly gaining ground. Public opinion still saw the Park Service—shades of "Ranger Rick" and "Jellystone Park"—as the good guys, and Barbee was on excellent terms with both his own rank and file and almost all the important politicians of the Northern Rockies.

And so it was that the Yellowstone Nation sailed into the summer of 1988 with serene confidence, some would say with supreme arrogance—the hubris of science combined with the sanction of the popular culture and the mandate of federal power. "We didn't think about fires at all," Barbee recalled. "Or at least not as anything serious." By August, however, the genial superintendent—the very image of the kindly camp counselor with his broad shoulders, ruddy face, and glint of gold cap in his smile—would be branded a villain and pilloried in editorials and cartoons across the land as the man who was "destroying Yellowstone."

He was a most unlikely candidate for the black hat. In the rumor-filled world of the Yellowstone Nation, few had a hard word against him. Maybe he was too easy-going. Maybe, it was said, he allowed himself to be pushed around too much by "the biologists," the "folks up at Research"—the scientists of Yellowstone Park's Research Division, that is, the professional class of wildland managers that outlasted all superintendents. The scientists, it was whispered, always anonymously, were the people who *really* ran the park, who set policy unchecked by independent peer review or other balances of power.

Yet Bob Barbee's affability masked a shrewd political instinct. Not by chance had he climbed through Park Service ranks to the top field position in the country. Like Dan Sholly in the Ranger Division and Don Despain in the Research Division, Bob Barbee was the cream of the crop, part of the elite. He knew where the true power was. He understood the politics of the office: how to juggle competing interests, who to stroke and who to scold, when to compromise, when to take a stand. He knew the mission.

Mission is a key word in the Park Service lexicon. Rangers go out to "complete the mission." Supporting natural regulation was "part of the mission." The mission of the national parks, in the words of the 1916 National Park Service Act, is to preserve the scenery, natural and historic objects, and wildlife "in such manner and by such means as will leave them unimpaired for the enjoyment of future generations." Yet does preserving the park conflict with the 1872 Yellowstone Park Act, mandating that Yellowstone should be for the "benefit and enjoyment of the people"? If "preserving" the park means, for example, allowing nature's ugly or frightening moments—fires, avalanches and mudslides, herds of starving animals in a tough winter, or the sight of a grizzly bear, as Barbee put it, "munching and crunching on a pretty little elk"—do these events conflict with the people's pleasure? Does natural regulation in fact serve future generations?

Those were the questions—in effect, the constitutional tensions—of the wildland democracy called Yellowstone. Most of the time, a compromise could be found. Barbee was a politician and compromise is the politician's business. But at the beginning of the fire season, in June, Barbee didn't see the need for compromise. Later, by the time Dan Sholly got back from the Calfee Meadow burnover, Yellowstone would be ready to compromise and fight a fire or two. In June, however, there didn't seem to be any pressing reason to abandon the mission, to abandon natural fire.

"Yes," Barbee said, "the West was in a drought. But Yellowstone is wetter than many parts of the West. April and May had seen precipitation well above normal. June was dry, but my experts predicted that the summer rains would come, as they always had, bringing 2 to 4 inches of precipitation. We expected an active but not highly unusual fire season—10,000, 20,000, maybe 40,000 acres at most. The forests needed fire. Of course, if we had known what was going to happen we would have done things differently."

38   Yet by the time Yellowstone Park officials came to fully accept that this was no ordinary fire season, it was too late. Some have argued that Yellowstone could have hit every single fire hard and early and there would have been no tremendous conflagrations. But that was not the mission. Others have argued that given the conditions in the region, by June it already was too late. For already something was stirring deep in the backcountry. Already that familiar force, as ominous and as common as a snake in the garden, had slid into their little piece of paradise.

III

Storm Creek

Fire is full of surprises. Fire management largely is about how not to get surprised, and what to do when the unexpected occurs. Letting a fire burn not only is seen by many as good for the ecosystem but also as the best way to manage a blaze. Firefighters must answer to budget-conscious superiors. The cheapest way to fight a fire often is to allow it to burn itself out against a natural boundary—a river, say, or a marsh, or a rocky ridge. In firefighter's lingo, this is the tactic of confinement. The fire is deemed "confined" if it is burning within a demarcated area; it is monitored—observed by foot patrols, planes, or lookout towers—but no direct action is taken against it. Confinement often works. The problem is, as long as a fire burns it retains its capacity for surprise and mischief.

Fire's great ally is the wind. A sudden gust or shift in direction can turn a placid little fire into a roaring menace, blowing it "out of confinement," across the river, over the marsh, beyond the ridge, into town. Every fire has a personality, a character shaped by circumstances of birth and environment. Some are mad dogs from the moment they're born.

40   Others are dull and plodding, predictable unto the day of
death, done in by humans on occasion but more often
defeated by weightier enemies: rain and snow.

Storm Creek was a secret agent among fires, a trickster
fire, unobtrusive and unremarkable for most of its life. It also
was the earliest major fire. It began with a lightning strike in a
stand of lodgepole pine above Storm Creek, a remote stream
feeding into the Stillwater River in the Custer National Forest,
about 16 miles north of Yellowstone Park and Cooke City. It
was not reported until June 19, when a mechanic test-flying a
Cessna out of Billings saw smoke over the Stillwater drainage.
Later, an outfitter's eyewitness report, confirmed by a light-
ning-detection system in Billings, pushed the date back by five
days, establishing June 14 as the start of the Storm Creek
blaze and the true beginning of the Yellowstone fire season.

The fire was in the Beartooth District of the Custer
National Forest. The district headquarters was in Red Lodge,
Montana, a colorful town of two-thousand souls at the foot of
the Beartooth Highway, a spectacular roadway that climbs to
11,000 feet in the Beartooth Mountains and on to Cooke City
and the park. When news of the fire came in to the Red
Lodge station, Fire Management Officer George Weldon
ordered an observation plane and took a look for himself.
Storm Creek indeed was burning in the Beartooth District of
the national forest, but it was inside a subunit of the area
called the Absaroka-Beartooth Wilderness. In forest-manage-
ment terms, the area was similar to Yellowstone Park. A
wilderness unit inside a national forest is preserved from
such development activities as logging and mining, and a fire
inside it is approached in the same way as a fire inside a
park. It is viewed as a natural part of the ecosystem and is
allowed to burn.

On June 19, Storm Creek was burning in 10 acres of
lodgepole pine. Weldon's estimate, backed up by the best fire-
behavior models of the U.S. Forest Service, was that Storm
Creek would burn about 600 acres under foreseeable conditions and 2,200 acres under severe conditions. The steep rocky
canyons of the Stillwater River formed a number of natural

firebreaks that were expected to confine the fire. By the end of the summer, however, Storm Creek would have swept across more than 100,000 acres and would be bearing down on Silver Gate and Cooke City.

If George Weldon had known what was coming no doubt he, like his counterparts at Yellowstone, would have hit the fires hard and early. But Weldon believed in the mission too. "For thousands of years lightning strikes have ignited fires which reduced heavy vegetation buildup and allowed for the renewal of forest and plant communities," he wrote in a public-information release attached to his fire report. "In addition to restoring the natural role of fire to the wilderness environment, prescribed fires have saved money. It costs a lot less to monitor a prescribed fire than to suppress a wildfire."

The main concern was that the fire would move about 5 miles north, beyond the wilderness boundary, and threaten campsites, private homes, and a chromium mine at the fringe of the national forest. It was hard to see how the fire could successfully run south. A generation of foresters had jokingly referred to the Custer National Forest as the "asbestos forest" because of its resistance to fire. A southward movement would take the fire deep into the forbidding Beartooth Mountains.

It isn't easy country. A narrow belt of steep forested hills follows the Stillwater drainage into the mountains, where sheer slopes rise to perennial snow fields and desolate peaks. On the other side of the peaks, tucked down into a crease of mountains along U.S. Highway 212, is the Northeast Entrance to Yellowstone Park and the gateway communities of Silver Gate and Cooke City. Fewer than seventy-five people lived in Silver Gate and Cooke City year-round, but to them the Great Spirit put the hamlets exactly in the right place. Bear and eagle, mountain lion, moose, and elk are their neighbors. The people are intensely proud of the area's wild beauty, its rough frontier history, and the independent lives they struggle to maintain. In the summer, the population swells to about 300, and thousands of visitors come through on their way to the park. Many in Silver Gate, Cooke City, and the surrounding

42   area make their living from the tourist trade, running motels, gift shops, restaurants, and other service-related businesses. A smaller percentage of the population earn their living in the old ways—cutting timber, working mine claims, ranching, and guiding hunters.

Around October, snow closes the road east and leaves the townspeople in deep isolation and uneasily dependent on their powerful neighbor to the west, the Yellowstone Nation. "People in Silver Gate and Cooke City had long been leery of the Park Service," Dan Sholly writes in *Guardians of Yellowstone*. "Though we provided—at the federal government's expense—law enforcement, fire protection, road maintenance, and medical services to their isolated communities, theirs had always been a 'we versus Big Brother' attitude. Those who depend on guided hunting and fishing, logging and mining activities are a highly independent breed, and many of them had never quite accepted the fact that nearby was a vast wilderness they couldn't simply utilize for themselves."

Sholly's remark accurately reflects the complicated relationship—a mixture of contempt and admiration—between the park and the gateway communities. The Park Service and its allies in the environmental movement see themselves as more sophisticated, better educated, and closer in touch with "the big picture"—the needs of the ecosystem—than the local rabble. "Most people have no idea how much the Park Service and the whole environmental herd loathe the gateway communities," said one Park Service employee with close ties to the gateway town of West Yellowstone. "They don't like the commercialism—the ticky-tacky souvenir shops, the restaurants and motels—and they especially don't like the locals constantly complaining about Park Service procedures and policies."

The Park Service attitude would translate into trouble when the subject became fire. For if, in June and early July, Barbee and Sholly and Despain and Weldon and a score of other experts saw no reason to abandon the policy of prescribed natural fire, the residents of the gateway communities did: it was just too damn dry.

"We tried warning Yellowstone Park time after time that those fires would spread," recalled Hayes Kirby, a tough-talking former Texan and owner of the Grizzly Lodge in Silver Gate. "But they just wouldn't listen. They said, 'Oh no, don't worry, it's just a bunch of small isolated fires. Everything is under control.' They thought they were better and smarter than us. The Forest Service wasn't much help either, and in any case they were overshadowed by Yellowstone Park. The park is a power unto itself."

Joan Humiston, a lifelong resident of the high country and owner of the Second Edition Cafe in Cooke City, remembers the early summer very well. "The rivers were way below normal in June," she said, "and the pine needles crackled underfoot, as if it was September. We all were nervous about fire." Just east of Cooke City, in the Clarks Fork Valley, Mike Evans was assistant ranch boss at the Griscomb Ranch. "Our hay was real low," he said, "and everything was drying up fast. It didn't take a genius to see we were heading into a severe fire season, but Yellowstone kept assuring us that everything was alright. Well, it wasn't alright."

For three weeks, as summer tourism picked up pace and area residents prepared for the big Fourth of July weekend, the region grew drier and Storm Creek smoldered, staying within the parameters sketched by George Weldon in his initial report. Then fire's ally paid a call. Winds gusting from 20 to 40 miles per hour swept out of the Stillwater Canyon. The fire ran more than 3 miles north through dry timber, past the rocky barrier where the fire analyst thought it would be confined, and spotted another mile ahead. Storm Creek had consumed 3,000 acres and now was threatening campsites and the chromium mine outside the wilderness boundary. By the evening of July 4, Weldon had completed the first Escaped Fire Situation Analysis (EFSA) of the Yellowstone season.

He was calling in the first of many Type I Incident Management Teams—the elite squads of the nation's top wildland firefighters—to handle the blaze. The team arrived in Red Lodge that night and decided on a plan of action. The fire would be allowed to burn on its southern perimeter, in the

44   wilderness, the side closer to Cooke City and the park. The strategy on the south side would continue to be one of *confinement*. On the north side, where it was moving out of the canyon, the strategy would shift to *containment*. Containment is a tactical step up from confinement. Whereas confinement allows a fire to burn with no suppression action taken against it—and with luck burn itself out against a natural barrier—containment means that firefighters will take *indirect* suppression action against the blaze. They would not go head-to-head with the fire—that tactic is another step up the suppression ladder and a dangerous business indeed—but would create "fire lines" of stripped earth beside burned-out woods to check the spread of the blaze. The language of these tactical shifts, clearly understood by firefighters, would be a source of much public confusion in the coming weeks.

With Weldon's July 4 EFSA, the game changed—if not yet inside the Yellowstone Nation then just a few miles north of its border. Storm Creek technically was not Yellowstone Park's fire. But like several other major fires of the season—in a terrific stroke of ill fortune for park officials—Storm Creek would move into their territory. Within ten days, with the burnover at Calfee Meadow under similar conditions, the game would change for Yellowstone Park too.

IV

Mammoth Hot Springs

In the days following Dan Sholly's escape from the firestorm at Calfee Meadow a number of early conflicts came to a head, establishing a pattern of mistrust and maneuvering that would severely undermine firefighting efforts. The source of these early conflicts was natural fire and the U.S. Forest Service. Forest supervisors—Barbee's counterparts—did not like natural fire and would not "accept" fires from Yellowstone Park, despite cooperative agreements that provided for fires to cross boundaries; that is, they refused to allow prescribed natural fires to burn from the park into their forests. Tension between the Park Service and Forest Service was nothing new, as indicated by their conflicting missions of preservation and multiple use. "The historic relationship between the two agencies," Pyne notes in *Fire in America*, "has been one of wariness, rivalry, and cooperation."

One of the characteristics of the Yellowstone fires was the speed with which crisis situations would shift from one wildland locale to another, frustrating the attempts of Park Service and Forest Service officials to establish a central command structure over the entire fire complex. At mid-July, the main problems were on Yellowstone's south and east flanks. Yellowstone's initial Fire Situation Analysis of the Falls fire, one mile north of the park's southern boundary with the Targhee National Forest, had determined that the blaze was natural and should be allowed to burn. But Targhee National Forest Supervisor John Burns was adamantly opposed to fire in his woods. The drought, wind, and fuel conditions bothered the Forest Service team. "We felt pretty certain that 1988 would be the worst fire year that we'd ever seen," Burns told Forest Service researcher David Peterson in an unpublished oral history of the fires. "We notified our rangers not to even think about anything other than an immediate suppression strategy on fires, because the potential was so great for them to get away."

On July 13, a letter from Burns to Barbee had arrived at the superintendent's office. In it, Burns wrote that although "in previous years with normal weather patterns, we would have been able to accept natural fires from Yellowstone National Park," this year they would accept no fires. Citing drought conditions and anticipated high winds during the summer, Burns noted that "other forests within our region are also experiencing similar conditions . . . consequently, our resources for controlling fires are in heavy demand at this time and are expected to be committed throughout [the] season."

Burns was anticipating trouble and moving to keep his resources—firefighters and equipment—close to home. This was a violation of the spirit if not the letter of wildland fire-fighting, which called on the various agencies to share resources. At Yellowstone, Barbee could not have missed the bureaucratic signal: don't count on us for help.

In addition, Burns had just forced Yellowstone's hand on the Falls fire. His refusal to allow the fire on the Targhee National Forest had triggered an interagency policy clause that called for Yellowstone to stop the fire at the park boundary. It works like this: if an administrative entity (Targhee National Forest) decides that a fire (Falls) does not meet its "criteria" (fire size and behavior, weather conditions, "values-at-risk" such as timber lots, watersheds, structures, or "visual resources") for accepting a natural fire, then it is the responsibility of the entity on which the fire originated (Yellowstone National Park) to take suppression measures to keep the fire from crossing the shared boundary. In plain English: we don't want it; you stop it.

Suppression action inside Yellowstone National Park began with an attack on the Falls fire on July 17. Fires had been burning in the Yellowstone area for over a month, since the ignition of the Storm Creek fire on June 14. Inside Yellowstone Park, fires had been burning for three weeks, since the start of the June 23 Shoshone blaze. The Yellowstone Fire Committee, meeting twice daily since Dan Sholly's shelter deployment on the Clover fire, had put out its first call for help and was mobilizing its own troops.

The call for aid was directed to the Boise Inter-Agency Fire Center (BIFC, pronounced "biff-see"), the national coordinating center for wildland firefighting. Established in 1965 to bring together state and federal firefighting agencies, BIFC's 55-acre site beside the Boise International Airport, in southwest Idaho, housed a high-tech military machine for the war against fire. BIFC gathers intelligence on fires from satellites and planes and detection systems, maintains a national radio network, keeps tabs on thousands of qualified firefighters all over the country, constantly trains personnel, is able to move hundreds of firefighters from its own airstrips at an hour's notice, and serves as a central base for determining which fires will have priority and receive critical resources. Contemporary wildland firefighting is a joint effort and BIFC was staffed by fire-management personnel from six federal agencies: the Department of Interior's Bureau of Land Management, Bureau of Indian Affairs, National Park Service, and Fish and Wildlife Service; the Department of Agriculture's U.S. Forest Service; and the Department of Commerce's National Weather Service.

It was still early in the fire season and BIFC's resources

48   were ample. Yellowstone's assistant fire-management officer, an easygoing, quietly competent ranger named Phil Perkins, requested one of the National Park Service's top firefighting crews—the Alpine Hotshots. "Hotshot crews" are the full-time elite of the firefighting profession. While the Alpine Hotshots were on the way, Perkins organized his own twenty-person crew from park personnel. On the morning of July 17, both crews were in place on the south flank of the Falls fire. Their mission was to keep every spark off the Targhee National Forest.

Right away, a problem developed. Running parallel to Yellowstone's boundary, at some points no more than 25 yards south of the line, was a Forest Service road. Perkins needed to conduct a "backfiring" operation to stop the Falls fire from moving on to the Targhee. His plan was to make an indirect attack on the fire, setting his own "backfire" from a safe, defendable "control line," and removing, by the use of his own fire, the fuel below the south flank of Falls.

The premise of a backfire is simple but critical: fight fire with fire; eliminate the fuel and the fire has nowhere to go. The logic of the situation pointed to using the Forest Service road as the control line. It would speed the entry of firefighters and equipment, save time and money, and contribute to firefighter safety. But the Targhee National Forest refused to give Yellowstone National Park permission to use the road.

Targhee officials would later claim they hadn't understood the request and would have approved it. That explanation is highly disingenuous. Forest supervisors were hardening their stands and were fixed, above all, on protecting their own turf. There was also the "visual resource" to consider: if Yellowstone was forced to backfire off its own boundary, the pretty forest scene north of the Forest Service road would be maintained.

The Yellowstone crew and the Alpine Hotshots trudged in along the narrow trail and set to work. The trail would have to be widened and cleared down to mineral soil, dead trees (called "snags") removed, and overhanging branches dropped along a 2-mile path to secure the control line. A day later, the job was done and the Yellowstone crew spread out along the line to guard against spot fires while the Hotshots moved along the trail setting fire to trees and brush with hand-held "drip torches," oilcan-like devices filled with a mixture of diesel fuel and gasoline. The flames swept quickly up into the dry trees and soon the fire began to creep north. The mission was a success: the zone of dry fuel between the Falls fire and the Targhee National Forest had been eliminated. But the actions of the Targhee officials left a sour taste in the mouths of Park Service management and firefighters. The storm clouds of distrust were gathering.

Along the park's eastern border with the Shoshone National Forest a roughly parallel situation was simultaneously developing. Here the border was not a map line drawn through the forest but a steep hydrographic divide—the high ridge of the Absaroka Range.

West of the ridge, inside the park, the Clover and Mist fires were exhibiting the unusual behavior that had been brought home so forcefully to Dan Sholly as he pondered his fate in the shake-and-bake fire shelter in Calfee Meadow. As noted earlier, the July 14 Fire Committee decision on Clover and Mist, however, was to let them burn. The committee was confident that, given the weather records, it would rain in the coming weeks. Certainly the blazes wouldn't blow over the ridge or up the long creek drainages onto the Shoshone National Forest. And even if they did creep across, that part of the Shoshone National Forest was a wild area, the North Absaroka Wilderness, in which lightning-caused fires were allowed to burn, provided they were "in prescription."

Prescription: it was the devil in the details of what constitutes a natural fire and when that fire will be defined as "escaped" and declared a "wildfire." A "wildfire declaration" was needed to trigger suppression activities; according to some fire officials, a wildfire declaration *compels* suppression actions. Unfortunately, Yellowstone Park and each national forest in the Greater Yellowstone Area had its own fire-management plan and its own definition of what constituted escape from prescription.

The fire plans of Yellowstone National Park and the Shoshone National Forest were contradictory. They represented, in the words of an official interagency review conducted in the winter of 1988, the two "extremes of the spectrum of fire prescriptions in the same ecosystem."

In the Shoshone National Forest, a fire was automatically out of prescription and declared a wildfire when it exceeded 1,000 acres, when moisture levels in the fuels dropped below a certain percentage, or when the fire moved out of defined areas. Yellowstone had no preestablished "circuit breakers" based on size or climate that would kick in a wildfire declaration. The decision of when to declare a fire out of prescription was left to the Yellowstone Fire Committee, a group with a strong commitment to natural regulation and natural fire. Thus, while the Clover fire was still burning *in prescription* at over 1,000 acres inside the park on July 14, it already had *exceeded prescription limits* for fire on the other side of the ridge in the Shoshone National Forest.

According to park officials, on July 14 the fire-management officer of the Shoshone National Forest told Yellowstone fire managers by telephone that the Shoshone would accept the Clover and Mist fires on to the North Absaroka Wilderness. So Yellowstone let the fires burn and concentrated their efforts on mobilizing resources to fight the Falls, Shoshone, and Red fires in the south, and the Fan fire in the north. A week later, the Shoshone Forest supervisor reversed that decision, telling Barbee that the Shoshone would not, in fact, accept the fires. Shoshone had done a complete turnaround. Meanwhile, Clover and Mist had grown from 5,000 to 12,000 acres. The probability of establishing control over the fires had radically slipped during the week.

The Shoshone National Forest has a slightly different version of the events from July 14 to 21. According to a dissenting note attached to the official federal review of the fires, the Shoshone fire-management officer had "explained that if the fires should reach the boundary, the forest would take suppression action. There was a Park misunderstanding where they thought the Forest would be accepting a prescribed fire." By July 15, the Shoshone had begun aerial monitoring of the fires, in preparation for the revised visit of Vice-President Bush—now excluding Yellowstone Park—and out of concern that Clover was approaching its boundary. From July 17 to 24, according to the Shoshone's account, there were "numerous discussions" between the fire-management officer and park officials. The fire-management officer "gave his personal and professional opinion about the fire situation. He clearly stated that any decisions to accept or reject any fires as a prescribed fire were not his, but the Forest Supervisor's." But where was the forest supervisor?

Forest Supervisor Stephen Mealey was doing what any sensible federal employee would do in the same situation—he was out fishing with his boss, the vice-president of the United States. His subordinate, overburdened with the twin duties of investigating a fatal helicopter crash in the Big Horn Mountains and monitoring the fires near the Shoshone National Forest, was left holding the bag.

As we will see, the forest supervisor's July 21 reversal was only the beginning of a long series of tactical and policy conflicts over Clover and related fires. And while the great blazes of Yellowstone's eastern sector certainly are not the fault of any one person, Yellowstone aficionados may long wonder what would have happened if Forest Supervisor Mealey had refused to accept the Clover fire on July 15, rather than on July 21. Perhaps he would have forced the park's hand (as Targhee Supervisor Burns had), obliging it to hit Clover the way it had hit Falls—early, and with a backfire to limit damage—before it was too late.

Eight miles south of the park, more trouble was brewing. On July 11, a fire-management officer flying a reconnaissance mission over the Teton Wilderness spotted four trees burning northwest of Enos Lake, near Mink Creek. The Teton Wilderness is part of the Bridger-Teton National Forest, covering 3.4

52   million acres in Wyoming. Like the Targhee National Forest to
the west and Shoshone National Forest to the east and north,
the Teton Forest includes preserved wilderness areas and
regions open to oil and gas exploration, mineral extraction,
timber harvest, motorized recreational vehicles, and hunting.
Outfitter operations are extensive and the area is dotted with
wealthy retreats. The guides and the upper-income ranch
owners were politically well connected and formed two powerful constituencies in the world of the national forest. As the
fire gained strength, and new starts multiplied, these constituencies would exert considerable influence on Bridger-
Teton Forest Supervisor Brian Stout.

Stout had decided to classify Mink as a prescribed natural fire. It was a risky call. The 1981 Teton Wilderness Fire-
Management Plan included incorporating "the natural forces
of fire" into the wild area, but weather and fuel conditions
were dangerously close to exceeding the plan's prescription
limits. Twenty-four hours later, aerial reconnaissance put the
Mink fire at 3 to 4 acres and spreading slowly northeast,
toward Yellowstone Park. By July 13, a strong wind was
rolling in and the fire was gaining momentum. By July 14, it
had exceeded 1,000 acres. Stout and his fire-management
team reassessed the situation. They decided to reverse the
three-day-old decision and reclassify Mink as a wildfire.

Stout's main concern was that the lower flank of the
Mink fire would move south into a "blowdown," a 15,000-acre
area of jackstraw timber blown down in 1987 by a rare high-
altitude tornado. Mink had to be kept out of the blowdown at
all costs. Fire in all that dead wood could touch off a catas-
trophic inferno that might sweep down on the homes and
ranches in the Buffalo Valley, 9 miles south of the blaze. An
early morning overflight of Mink on July 15 indicated that
indeed it was moving south and had grown to over 6,000
acres. By afternoon it had progressed an additional mile and
was nearing the fringe of the blowdown. By midnight, it was
at 9,000 acres.

Stout completed an Escaped Fire Situation Analysis and

called in a Type I overhead team led by Incident Commander Dale Jarrell. Their strategy was to attempt to contain Mink on its south flank with suppression actions and let it burn to the north, toward Yellowstone Park. Jarrell ordered ten crews (200 firefighters), four helicopters, support equipment, and personnel from the Forest Service's Region One headquarters in Missoula, Montana. In Idaho, BIFC put an infrared flight over Mink at midnight, July 16. The blaze had increased to 12,000 acres. The next morning, Jarrell established his incident command post and set two "spike camps"—forward firefighting posts—near the southern perimeter of the blaze.

In addition to Mink itself, there was another matter to consider. This was Mr. Griz's country—prime grizzly bear habitat. *Ursus horribilis*, usually weighing in at well over 500 pounds and with 4-inch claws, enjoyed camp food and occasionally was known to crunch an unlucky citizen. Strict food and garbage storage methods would have to be enforced, so as not to tempt bears into camp—no easy task with 200 people and a nasty fire in the neighborhood. All food, garbage, personnel, and tools were transported in and out of the spike camps by helicopter.

Keeping a wary lookout for bears, Jarrell's crews set about using portable water pumps to wet the edge of meadows and began to backfire areas south of the fire's perimeter. Stout—attempting to stay within "light-on-the-land" wilderness-use guidelines—approved only a limited use of chain saws to help build fire lines. Helicopters conducted "bucket drops" by dipping collapsible 100-gallon fiberglass-and-rubber buckets at the end of long cables into lakes or rivers then lifting to water the fire area and dropping the water on "hot spots" inside the Mink perimeter. But the effort did not go well. Mink continued backing south and began to exhibit high-intensity behavior: crown fires in the treetops and frequent spotting over the lines.

By July 18, smoke was pouring into Cody and the townspeople were up in arms. A new infrared flight put the fire at

54   14,000 acres, with the south perimeter closing on the blowdown and the north flank 3 miles from Yellowstone's border.
Wyoming's powerful U.S. senators, Malcolm Wallop and Alan
Simpson, were on the line to Barbee and Stout, pressing for
information and action. Wallop was planning a trip to the fire
to see for himself. And the media were beginning to hammer
on the door. For over a month it had been a local story—the
Montana, Wyoming, and Idaho media had been on top of it
from the beginning, closely following developments. Now it
was becoming a regional story, with Denver and Salt Lake City
sending reporters and film crews to the fires.

Meanwhile, more crews were arriving to fight the Mink
fire. Stout posted an information officer to Cody to keep the
community in touch with developments and contacted Yellowstone fire officials. The northern head of the fire was now
on the Yellowstone Park boundary. Would the park accept the
Mink fire? Barbee indicated it would. If Barbee had decided
not to accept the blaze, as others had decided not to accept *his*
fires, Stout would have been forced to concentrate major
resources to the north. In what was perhaps a gesture of
thanks, Stout amended his EFSA to include protection of Yellowstone Meadows, on the park boundary, as a second priority. Water hose and personnel were dropped into the meadow.

The blowdown remained the top priority. Mink was spotting into it. Jarrell and his operations staff came up with a
high-risk manuever. They wanted to put control lines through
a sliver of the blowdown and burn out 900 acres. Since Mink
was already burning in the fringe of the blowdown, this was a
direct-attack operation, unlike Phil Perkins' indirect attack on
the Falls fire. In a direct attack, with control lines close to the
fire, the tactic is called a "burnout," rather than a backfire.
The danger was that the burnout could sweep right over the
control lines and into the main body of the blowdown. Stout
approved the operation.

Jarrell's crews prepared the control lines. Helicopter
teams armed the aerial ignition devices that would drop fireballs into the burnout zone. By noon on July 20, everything

was coming together. Stout and Jarrell lifted off in a helicopter to observe the proceedings. It was, Stout later told Forest Service researcher David Peterson, "the scariest burnout
I've seen in my life," and a dangerous career moment as well.
If the burnout failed and the blowdown torched off, heads
would roll. Just as the operation was about to begin, a Salt
Lake City television crew was discovered inside the burnout
area. They had bypassed the information officers and hired an
outfitter to take them to the fire. The burnout was suspended
until the media were escorted from the area. Then the helicopters swept through, dropping the fireballs. Smoke started
to roil below, climbing 10,000 . . . 20,000 . . . 30,000 feet.

Radio reports flowed in from the lines, Stout later
recalled. The fire was everywhere. The heat, the smoke was
overwhelming. Soon, however, came the good news. Spot fires
were minimal. The lines were holding. The south flank was
secure.

Stout and Jarrell had succeeded. Phil Perkins also had been
victorious on the Falls fire. But on an important policy level,
four of the principal managers of the Greater Yellowstone
Area—Park Superintendent Barbee and Forest Supervisors
Burns, Mealey, and Stout—were marching off in different
directions. During the ten-day period from July 11 to July 21,
the four officials had embarked on a series of uncoordinated
policy moves that underscores the often conflicting values of
the National Park Service and Forest Service, and goes a long
way toward explaining the confusion that was settling in
among both the public and the growing number of firefighters
on the lines.

The national forests surrounding Yellowstone Park, after
a gesture in the direction of natural regulation, had reverted
to their traditional hostility to fire and were moving quickly
toward aggressive firefighting. At the Targhee National Forest,
John Burns had made his position clear: he wanted no fires
on his property. On the Shoshone National Forest, Steve
Mealey had initially signaled that he would accept the Clover

56 and Mist fires, or at least his subordinates had, only to reverse—or at best "clarify"—his position a week later, when the fires nearing his border had tripled in size. On the Bridger-Teton National Forest, Brian Stout smelled trouble and within three days had shifted the classification of Mink from "prescribed natural fire" to "wildfire"—a shift that triggered the suppression effort.

At Mammoth Hot Springs, although officials remained committed to the philosophy of natural regulation, biologist Don Despain was starting to lose the argument for letting the fires burn. Politics was taking over biology. Barbee had approved suppression action against the Falls fire and the human-caused Narrows fire, and by July 22 he would be setting in motion attack plans against two more major blazes.

One afternoon shortly after the successful operation on the Falls fire, three members of the Yellowstone Fire Committee—Dan Sholly, Don Despain, and Fire Management Officer Terry Danforth—lifted off the Mammoth helipad with pilot Curt Wainwright. They wanted a firsthand look at the Shoshone fire, burning roughly 12 miles north of the Falls blaze.

All four men remember it as a time of mixed emotions. They believed that natural fire played an overwhelmingly positive role in the ecosystem. Their mission was to preserve the policy. But they sensed they were getting into a whole new game. The fires were starting to outstrip their projections, and some had begun to threaten structures and roads in and around the park. What was the mission now? To put the fires out, or to let them burn? Where was that invisible line between protecting the policy for the future and letting nature take its course in the present? Had they already unwittingly crossed that line, bringing a political if not an environmental disaster down on the world they cherished?

Don Despain had been around a long time. "I've argued with lots of superintendents," he recalled with a chuckle. " 'Don,' they all would say, 'you've got to put some out to let others burn. That's politics. That's reality.' They would joke at the Fire Committee meetings, 'Don't ask Despain for his vote, he's for lettin' 'em all burn, he's for lettin' 'em all go.' Sure I was," Despain adds. "That was good science. That was good biology. That was reality too—the reality of nature."

The wind was picking up and Shoshone fire monitors a few hours earlier had reported crown fires and spotting up to a half-mile ahead of the main front. The fire was making its first major run.

"Despain was interested in that one," remembers pilot Curt Wainwright. "It was nearing an area of young lodgepole pine that had been burned over in the late 1940s. Those young trees probably would slow the fire's movement."

"It was thrilling, in a way," says Don Despain. "The world knows so little about fire ecology, and because we already had so much baseline biological data on Yellowstone this was the chance of a lifetime, a truly historic opportunity." If only they would let the fires burn!

"Shoshone had moved about 2 miles northeast that day," Fire Management Officer Terry Danforth remembers. "With the cooling afternoon temperatures and rise in humidity, it should have been slowing down just south of the Continental Divide, a line of decision for us. Any future runs or spotting would pose a direct threat to the Grant Village visitor complex."

The helicopter closed in on the blaze. They peered down at the wave of flames rolling into the young pines. The smoke column towered above them. Embers and ash swirled through the air. "Sholly says to me, 'Can you get any closer?' " Wainwright recalled. "I veered right, keeping the craft about 500 yards beyond the inferno. It was a beautiful thing, magnificent really. We were tiny beside it. But Despain wasn't watching the heights of the fire. His eyes were fixed on the ground. The fire was marching right through the young growth, hardly slowing at all! That was one for the books."

Wainwright had steered the ship to within 400 yards of the fire. He was flying at 1,500 feet. "Can you get any closer?" Sholly asked again. But something weird was happening.

58   Wainwright struggled to bring the helicopter in a tight turn past the fire's face. "The ship wasn't responding," Wainwright remembers.

They were losing altitude. The fire was sucking them in. A thermal wave jolted the craft. Sholly turned to Wainwright, asking, "Can you . . . "

Wainwright jerked a finger at the altimeter.

Sholly looked at the rapidly dropping numbers, then out the window, then at Wainwright. "Curt," he said quietly, "you do what you have to do."

Wainwright pitched out hard, down, away from the blaze, and headed for home.

V

Snake River

Late in the evening of July 21
the phone rang in the home of Dave Poncin, fire staff officer
of the Nez Perce National Forest in Grangeville, Idaho, and
one of only eighteen Type I incident commanders in the country. A U.S. Forest Service regional coordinator was on the
other end of the line. Unusual fire behavior and threats to
structures had led Yellowstone to reassess the situation
regarding the Shoshone, Falls, and Red fires in the southern
end of the park. The status of the fires had been changed from
prescribed natural burns to wildfires. The Park Service had
requested a Type I team. The three fires to be placed under
Poncin's command would be given a new designation: the
Snake River Complex.

Poncin was surprised by the call. Yellowstone, with its
natural fire policy and a corps of highly trained firefighters,
rarely went outside the park for help. Poncin began to work
the phone. His ten-person Incident Management Team, fighting fires since late May, had been rotated off the fire lines for a
rest a week earlier. They were spread out over Idaho and Montana. The Forest Service dispatched planes and the team ren-

60   dezvoused in Missoula in the predawn hours of July 22. They flew to the gateway community of West Yellowstone in time for breakfast. Then they drove 50 miles to Mammoth for a meeting with the Fire Committee.

Poncin, fifty-one years old, with thirty years of firefighting experience under his belt, had great affection for Yellowstone Park. He had been a Forest Service ranger in the gateway community of Gardiner in the late 1970s. With his children and Montana-born wife, he had spent countless bucolic hours in Wonderland.

Still, as he would later recount, something was troubling him. Type I teams almost always were asked to *suppress* fires; officially, a "wildfire" declaration was supposed to initiate suppression activities. But Yellowstone apparently was allowing the flanks of some of the fires to run as prescribed natural fires, while declaring the Snake River Complex a wildfire.

It was confusing. The July 20 Escaped Fire Situation Analysis on the Shoshone fire, for instance, called for containing the east and north flanks by reducing fuels and burning out along highway boundaries, but allowed other flanks to continue as prescribed natural burns. Yet if they were in a wildfire situation and a containment mode—both of which called for suppression action—how could the blaze still be classified as a "prescribed natural fire"?

"It was the wrong classification," a Yellowstone official later conceded. "Our reasoning was that we didn't want to construct fire lines over the whole complex. But clearly we used the wrong term. At the time, we were just too busy to worry about it."

Then there were the light-hand, or minimum-impact suppression, tactics. What, exactly, did the park mean by this? When could bulldozers be used? What about chain saws? What about the use of explosives to quickly build fire lines? A fire line by definition must be cut through grass and duff, down to mineral soil. Was it really a sensible light-on-the-land tactic to use only hand tools—shovels, rakes, and the double-bladed hoe-and-axe called a "pulaski"—to construct a fire line,

when dozers, chain saws, and explosives could do the job
much quicker?

On these matters, the official review of the Snake River fires would note that the "concern expressed by [Poncin] in not 'putting the fires out' apparently was felt by other ICs [incident commanders] assigned to Yellowstone National Park in 1988." Although Poncin and later ICs "did successfully adjust to a limited-objective strategy, this type of strategy is not universally accepted within the fire-management community." The confusion over light-hand tactics was also noted. "There are too many differing definitions of what it means in suppression," the review concluded.

Poncin's concerns were addressed at the July 22 Fire Committee meeting at 10 A.M. The park clearly defined what, in this particular situation, it wanted from fire management and what tactics were sanctioned. The top priority was the protection of lives and structures at the Grant Village tourist center on Yellowstone Lake—occupied by some 4,000 visitors and staff—as well as the West Thumb visitor area, the Lewis Lake campground, and the Montana Power electrical substation and transmission lines. Poncin's other directives were to keep the Falls fire in the park, and to prevent the Shoshone and Red fires from severing the highway running from the south entrance to Grant Village and on to the "Grand Loop" road, Yellowstone's main economic artery. Bulldozer use was limited. "We strongly felt that the dozers cut much deeper and did much more damage than lines made with shovels and rakes," Barbee said later, "so we limited their use. Chain saws were approved, however. We were still struggling to find the right balance between fighting the fires on the one hand, and preserving and protecting the land." Poncin, for his part, did not disagree with the tactics and remembers being "comfortable with the briefing."

For Bob Barbee, the stakes had just gotten higher. Would they have to close Grant Village? Barbee wanted to give IC Poncin as free a hand as possible, although in the superintendent's opinion no one knew the terrain and weather patterns as well as his own people. A hasty evacuation of Grant would totally disrupt a major operation of Yellowstone's chief concessionaire, TW Services, and possibly deal a devastating blow to park visitation for the rest of the summer as the news flashed nationwide. A long closure of the South Entrance Road would be almost as bad. Barbee realized the fires were growing too big and too complicated for direct park control, yet he resisted surrendering Yellowstone's fate to outsiders, even to one as well-meaning as Dave Poncin.

Poncin drove down to Grant Village with his team. Yellowstone had given him grounds just north of Grant Village for his base camp, but he had few crews. First, he needed logistical support from BIFC. "It was obvious to me from the park briefing that there weren't enough crews and engines on hand to deal with a major confrontation with any of the fires in the Snake River Complex," Poncin recalls. "I wanted my base camp up and running, with at least 500 firefighters ready to work day and night. Second, to perform, the team needed up-to-the-minute intelligence. Fire Behavior Analyst John Krebs was a member of my overhead team. Krebs would fly the fires, prepare a new analysis, put scouts linked into the fire-command radio network out on the fires, and quickly build a support group of observers and behavior specialists. Next, he would prepare briefings on past, present, and future fire behavior, cautioning about hazards and helping shape fire strategy."

The team had worked together for years and the wheels of the operation were soon in motion. Crews were ordered from BIFC. Carpenters and electricians arrived and began to bang together the rough structures that would serve as command post and support offices for the Snake River firefighters. They were raising one of the makeshift cities of the fire world, complete with mess halls and medical units, latrines and light poles, showers and supply caches full of gear. Dozens of these boom towns, from tiny spike camps deep in grizzly territory to sprawling fortresses rumbling day and night with the sound of trucks, generators, and helicopters, would rise and fall in the Yellowstone area over the next eight weeks.

Meanwhile, Dave Poncin was doing some scouting of his own, taking a drive through Grant Village. "What I saw made my blood run cold," he remembers. "Immediately on entering the compound, I saw that logging debris from the initial clearing of Grant years earlier had been piled at the forest's edge. Debris and firewood also were stacked beside buildings. The structures all had cedar-shake roofs, the worst sort of roofing in terms of fire control. It was a disaster just waiting to happen. If the fire blew in that day, we were finished."

He steered carefully, slowly wending his way between tourists and buses and recreational vehicles. Yellowstone Lake gleamed at the edge of the development. The wind was picking up. Poncin drove down to the large boat ramp jutting out into the lake. They had decided to use it as a heliport. "I walked out on that ramp and turned to look at the village tucked into the trees at the side of the water," Poncin said. "Smoke was rolling in, and with it a dusting of ash. A strange kind of wind was blowing, possibly caused by a fire draft that was sucking the blaze forward. Although I couldn't see it, I knew Shoshone was on a run." That day the Shoshone fire moved to within 2 miles of the Grant Village highway junction.

Poncin walked back to his car. "I had made up my mind. Grant Village would have to be evacuated."

Before dawn on July 23, Park Service rangers and TW Services' personnel were knocking on doors and circulating through the Grant campgrounds. Everyone, without exception, would have to leave. Immediately.

Dawn came, and with it more fire crews. As the tour buses and cars exited along the congested road, Poncin's crews set to work. "My people were eager to go out and hit the fire where it lived," Poncin said, "but this was to be a defensive action. We would wait for the fire to come to us."

There was plenty to do. Poncin wanted a 150-foot cleared area around the perimeter of Grant Village, and all the buildings fireproofed. Pumper trucks and sprinkler systems bathed the structures in water and foam. Backhoes hauled away the piles of old debris. Sawyers armed with chain saws dropped

64   dead trees and cut off the lower limbs of live trees to deprive a possible ground fire of the "ladder fuels" it needed to climb into the thick canopy.

They were racing against time and the weather. Under normal weather conditions, a fire will "lay down" at night beneath cooler air and higher moisture levels. As the morning sun warms the forest, the cool air that has pooled over the terrain—called an "inversion"—begins to break. An inversion helps firefighters by slowing the burn, although the thick layer of smoke hinders them by reducing visibility. (First rule of firefighting: find the fire.) When the inversion lifts, the moisture level drops and the fire picks up speed, sometimes explosively. Inversions usually break around noon, leading to an active afternoon burning period, until evening brings cooler air and a rise in the moisture.

But odd things were starting to happen in the Yellowstone country. Sometimes the fires would not lay down at night. The moisture levels would remain stubbornly low. The rains would not come, and in their place a series of dry, gusty cold fronts would roll in. The blazes were beginning to head off the charts.

On the afternoon of July 23 the inversion lifted and the Shoshone and Red fires took off. "I was on the road outside Grant," Poncin remembers, "watching the Shoshone advance. Showers of embers, like wind-driven waves, were crossing the road." Poncin called for more crews and ordered the South Entrance Road temporarily closed.

Bob Barbee's fears were coming to life. The road closure meant that no visitors could enter the park from the south. The Grant Village evacuation was drawing national attention. With the wind gusting close to 20 miles per hour and Shoshone less than a half mile from the village, it seemed likely that Grant would go up in flames. And Dan Sholly had brought Barbee more bad news: a new fire, destined to be the biggest in Yellowstone's history, had entered the park from the west. It was pointed straight at Old Faithful.

VI

North Fork

$\mathbf{A}$t about the time the Shoshone fire was making its first major run toward Grant Village on the afternoon of July 22, Leland Owens was calling it a day. Owens and three other men had been cutting firewood on the North Fork drainage of Moose Creek Plateau, a western section of the Targhee National Forest, 250 yards from the boundary line with Yellowstone Park. It was a Friday afternoon and the men had been working hard. Although their permit technically allowed cutting only for personal use, they made a few extra dollars in the winter by selling the firewood. Times were tough and the Forest Service tended to look the other way when it came to a few individuals scratching out a living. The woodcutters had a beer and a smoke, and left. Later, firefighters would follow the fire's trail upwind and discover the cigarette butts and beer cans. The butts were from a generic brand of cigarettes sold at the nearby Fort Hunt Indian Reservation. Checking the Forest Service's logging registry for July 22, investigators turned up the names of Owens and the others. A single butt, found at the fire's point of origin, was traced to Owens by a saliva test. The wind had blown the

66   fire steadily away from the discarded cigarette, thereby preserving the evidence of the ignition of Yellowstone's largest and most controversial fire.

It was a good day for a fire, hot and dry, with a strong breeze. The parched grass kindled up fast and the southwest wind quickly blew the crackling flames into a vast stand of decaying lodgepole pine, much of it blown down by windstorms or killed by a pine-bark beetle infestation.

Within an hour, according to the federal review of the North Fork fire and records obtained from Yellowstone Park's Public Documents Center, the blaze had been spotted by a Forest Service employee, who called District Ranger Rodd Richardson. Richardson immediately ordered up a plane and smoke jumpers from West Yellowstone, 15 miles north of the blaze, and sent initial-attack ground forces to the area. The smoke jumpers circled the fire at 2:50 P.M., reporting it to be 75 acres wide and heading for the park. Jumping it was impossible: the winds were too high. From his office at the Island Park Ranger Station, 15 miles west of the ignition point, Richardson could see the smoke column rising. The fire, he later said, "was really ripping."

By 3:10 P.M. the first ground forces arrived from Island Park. The fire already had crossed into Yellowstone. By late afternoon, suppression forces included three pumper trucks, two bulldozers, a fifteen-person hand crew, and air tankers from West Yellowstone. They hit the fire hard, the dozers tearing away the forest on its flanks and moving toward the park, the hand crew crossing into Yellowstone. The fire was too hot and moving too fast for effective retardant bombing, but the small lead plane used to guide in the tankers reported spotting a half-mile ahead of the front, an observation later confirmed by infrared reconnaissance.

Richardson had contacted his opposite number in the national park, West Yellowstone District Ranger Joe Evans. The ranger was soon on the scene. The fire could be stopped, Richardson said, if the park would allow bulldozers to cross the boundary and cut a line around the head of the blaze.

Night would bring a drop in the wind. The fire would lay down. The terrain was relatively gentle. They could do it.

Evans, who had already discussed the situation with Dan Sholly, denied the request. Park policy was clear: no dozers in the wilderness. The deep slashes bulldozers made in the earth, Yellowstone officials argued, were not natural and took much longer to heal than the scars of fire, which were natural. Furthermore, Yellowstone's Madison Plateau, onto which the fire was spreading, was a sea of overaged pine. In the view of park biologists, the Madison Plateau needed a fire.

There was also the matter of firefighter safety. Evans and Richardson knew they were facing an extremely dangerous situation. The area was full of snags—the standing dead trees that could topple over with a whisper of wind. The fire was spotting. There were few safety zones to shelter firefighters. Evans said the hand crews could continue to work—during the night shift they would put out one spotter for every two crew members because of snag danger—but no dozers would be allowed over the border.

The "dozer decision" would become a major point of controversy, a defining moment in the public perception that Yellowstone Park had blindly wedded itself to the "let-burn" policy and was callously disregarding local concerns. North Fork, after all, was a human-caused fire. Yellowstone's policy called for the immediate suppression of human-caused fires.

But the policy also prohibited the use of heavy machinery in the wilderness. Barbee could have overruled that aspect of the policy, as he did in other cases, but he chose not to. By the time its rampage was finished, North Fork would hit Old Faithful, Canyon Village, Mammoth Hot Springs, threaten Tower Falls and West Yellowstone, burn over a half-million acres, and cost more than $36 million to fight.

Could a few dozer lines inside the park on July 22 have stopped it? The fire was spotting a half-mile into dry forest by the time the dozers arrived. Could the dozers and, say, fifteen twenty-person crews to man the fire lines and chase spots, plus aggressive retardant and bucket drops, have done the

68   job? Perhaps. And perhaps gusting winds would have trapped crews in a flaming crypt or under toppled snags. With 300 firefighters on the scene, strategy would have mandated clearing wide firebreaks and safety zones all over the Madison Plateau—mandated the use of bulldozers, in other words.

And where would the crews have come from? In fact, there were a couple hundred armed and ready firefighters close to North Fork. Dave Poncin's teams were deploying on Snake River, concentrating around Grant Village. The lakeside development was at that moment the top priority. It made no sense to Yellowstone officials to pull limited resources out of Grant to fight a remote fire in an overgrown area. Along with firefighter safety and policy proscriptions, the critical question of resource allocation was now coming into play.

As darkness came, Rodd Richardson hurried back to the ranger station and completed an Escaped Fire Situation Analysis. It was 9 P.M. North Fork was roughly 460 acres. The dozers, aided by the southwest wind, had tied off the portions of the fire inside the Targhee National Forest by driving wide lines around its flanks. The wind was herding the head of the blaze deeper into the park, where hand crews were at work. The EFSA, coordinated with park representative Evans, called for a "control" strategy—active suppression measures—on both forest and park lands. Dozers would work the Targhee side of the boundary line to protect tree plantations from a backing fire. Hand crews would build lines and conduct burnouts inside the park. But what they really needed was more firefighters. The EFSA requested a Type I Incident Management Team.

Technically, then, with its human-caused origin and the request for a Type I team, the North Fork fire was classified as a "wildfire" from the outset, a classification that called for suppression action. Yet what exactly *was* suppression action? Did it have to be "light on the land"? Could they use bulldozers? Chain saws? The definition seemed to shift from day to day, and from fire to fire. It appeared that the only thing Yellowstone Park wanted was to let the fires burn.

With conflicting information emerging from the various fires, mushrooming smoke columns on the horizon, and a smoky haze blotting out the pristine skies, a state of siege was beginning to settle over the gateway communities and isolated ranches of the Yellowstone area. Dark rumors, as from the deep paranoia of a war zone, were circulating: the park didn't want the fires suppressed at all; it was a plot, a radical environmentalist conspiracy to wipe the "tourist trap" gateway communities off the map and return Yellowstone to some pre-Columbian pipedream.

There was more than a grain of truth in local fears. Usually the contempt environmentalists feel for the year-round, working residents of the high country is kept strictly off the record. But in a remarkably frank moment after the fires, National Audubon Society board member Scott Reed told the Idaho Conservation League that "the greatest environmental disaster coming out of the Yellowstone Park fire was its failure to burn up West Yellowstone. In the fierce competition between Wyoming and Montana for the ugliest town, West Yellowstone is the easy winner. What a wonderful thing it would have been to reduce all that neon clutter and claptrap to ashes."

Even before the North Fork fire announced itself, another important new player was entering the picture. The fires had posted a clear increase in severity since Dan Sholly's shelter deployment a week earlier; firefighting resources were not arriving at a pace equal to fire spread; and all the political units in the area were backsliding into the First Law of Bureaucracy: when there's trouble, protect yourself.

On July 21, Barbee had called his boss, Denver-based Rocky Mountain Regional Director Lorraine Mintzmyer. They reviewed the overall situation. The Fan fire was kicking up in the north. Dave Poncin was on his way to Grant Village in the south. To the east, Clover and Mist had merged and the Shoshone Forest had refused to accept it—a wildfire declaration was imminent. Barbee had stated that he was prepared to

70 accept the Mink fire from the Bridger-Teton Forest, but the Targhee Forest had refused to accept the Falls fire. This was coordination? This was interagency cooperation? Something needed to be done to bring the entire situation under control.

On July 22, Mintzmyer called her Forest Service counterpart, Rocky Mountain Regional Forester Gary Cargill. The suppression effort, they agreed, was growing increasingly complex. After consulting with other regional officers, they decided to take the next step up the Incident Command System, calling in an Area Command group to coordinate suppression activities.

Area Command was a joint Forest Service–Park Service operation, led by two officers, one from each agency. They were to set up headquarters in West Yellowstone and serve as an overall coordinating body, controlling the flow of resources through the region, and providing central intelligence on fire behavior and weather trends. Area Commands for large fires had been set up before, but never for a region with so many overlapping jurisdictions—encompassing three states, twelve counties, six national forests, and two national parks.

Working in conjunction with the Greater Yellowstone Coordinating Committee (GYCC), a federal interagency group, Area Command would attempt to head off problems between park superintendents, forest supervisors, incident commanders, state and local authorities, and the gateway communities. The understaffed GYCC had been attempting to do this for years: Area Command was a kind of fire-emergency GYCC. Area Command would also attempt to coordinate the public-information campaign, but its central function was to get firefighting resources to the right places at the right times.

Although the separate incident commanders—Dave Poncin and Dale Jarrell, soon to be followed by a host of others—were free to make tactical decisions, Area Command would decide which fires had priority, and would make certain that a strategic decision on one fire did not conflict with a strategic decision on another fire. Of course, by controlling resources Area Command could in effect dictate strategy. That point was not lost on incident commanders.

On July 23, Troy Kurth, a veteran Forest Service fire-fighter, arrived at Yellowstone to serve as one of the two area commanders. Dan Sholly took him up for a helicopter overview of the situation. As Sholly recalled it, Kurth did not like what he was seeing. All over the park, the fires were moving steadily northeastward. Yet the business of the park seemed to be continuing as normal. The roads and camp-grounds were crowded with visitors. The smoke columns and burning woods seemed to be just another roadside attraction, along with the bison, the elk, and the geysers. If it didn't rain soon, Kurth would be faced with the prospect of moving thousands of firefighters and tons of equipment in along the same roads the visitors were using.

They flew over Old Faithful and surveyed the perimeter of the North Fork fire. The ragged edge of the blaze was seizing tree after tree, flames tracing up the trunks and exploding in the canopies. Its general movement was north, with an easterly bulge and a bit of backing fire to the south. On July 22 it had been at about 400 acres; twenty-four hours later, it was up over 1,000 acres. That was bad, Kurth believed. They hadn't even had a high wind day. One good gale and the fire would rip across the Madison Plateau and onto the structures in the Upper Geyser Basin around Old Faithful. Kurth told Sholly they should evacuate Old Faithful immediately.

"Evacuate Old Faithful!" Sholly remembers. "I was stunned. We had just evacuated Grant Village!" If Yellowstone was the crown of the National Park system, Old Faithful was the jewel in the crown. Closing it down would be the same as closing the park, admitting they were whipped. It would destroy park morale and ruin the regional economy. Sholly told Kurth he opposed the move. "I thought Kurth was overreacting," Sholly said. "He was new to the situation. North Fork was still at least 8 miles from the Upper Geyser Basin. That was a lot of distance for a fire to cover. I respected Kurth, but I felt I knew the country much better than him. The fire could wheel north. We could flank it at one of the small streams running off the plateau. We could do some burnouts. It could rain."

Kurth agreed to wait. Larry Caplinger was coming in

from the Stanislaus National Forest in California with his Type I Overhead Team to assume control of the North Fork fire. They would confer with him. By 6 P.M., Caplinger had arrived and took over from Targhee Ranger Rodd Richardson as incident commander. Sholly drove down to West Yellowstone for a meeting with Caplinger and Kurth. Caplinger outlined three options: let the fire burn and meet it at Old Faithful; encircle the blaze with a risky multi-million-dollar fire line; or use fire lines and backfires on the south and east flanks to keep the blaze away from Old Faithful, while letting it free-burn north across the Madison Plateau.

The decision was left to Barbee. "It was unacceptable," Barbee says, "to wait for the fire at Old Faithful. Circling it was too dangerous and too costly, with little chance of success given the wind patterns and timber moisture. So that left the third option. We would secure the south and east flanks, let it burn north, and pray for rain."

Barbee's refusal to allow dozers onto the Madison Plateau and the July 23 decision on North Fork underscore the three key elements that repeatedly influenced firefighting in Yellowstone: *safety*—incident commanders didn't want their crews at the head of potentially lethal blazes like North Fork; *policy*—although moving toward active battle of the blazes, Barbee wanted to maintain natural fire when and where it was possible to do so, both to lessen mechanized suppression damage, and to protect the policy for the future; and *resources*—there weren't enough crews in the area to safely manage all the fires, and so at the moment Grant Village remained the top priority.

Nature, however, soon stuck in its windy hand again and forced man to readjust his priorities. On July 24, driven by high winds, North Fork doubled its size and slid 4 miles closer to Old Faithful. Larry Caplinger relocated his command post from the Island Park Ranger Station to Old Faithful and called in structure protection units to help defend the development. Area Command placed a large order through BIFC for engines and firefighters. It was becoming apparent that they would need everything they could get their hands on. At Grant Village, Dave Poncin was braced for the Shoshone fire's assault. And over on the eastern boundary, a fire had moved onto the Shoshone National Forest. A new incident commander was on the scene. His name was Curt Bates and he was running into problems.

VII

Clover-Mist

At 3 A.M. on July 24, Incident Commander Curt Bates arrived at the Cody Armory and prepared to take over the Clover-Mist fire. Following the Shoshone Forest's refusal to accept the fire, Clover and Mist, which had burned together on July 22, had been declared a wildfire. Hours before the new incident commander's arrival, the southern flank of the Clover-Mist fire had moved out of the park and into the North Absaroka Wilderness area of the Shoshone Forest. The northern flank of the fire was burning east up the Miller Creek drainage, toward the Clarks Fork Valley. Retardant drops were ordered in an attempt to slow the fires while the crews were put in place.

Bates' assignment was to take over a complex comprised of seven fires, all of which began as prescribed natural fires in the northeast sector of Yellowstone Park: the Clover, Mist, Raven, Lovely, Shallow, Fern, and Sour fires. He designated the complex "Clover-Mist" after the two biggest blazes, was briefed by Yellowstone and Shoshone officials, and studied the latest Escaped Fire Situation Analysis.

The lanky veteran firefighter immediately recognized the

76   problem. The fire was burning at approximately 13,000 acres inside Yellowstone Park, with less than 200 acres slopping over into the Shoshone National Forest. Yet his orders, dictated in the EFSA, were to allow the fire to burn in the park, while quickly moving to suppress the "slopover" in the Shoshone Forest with light-hand tactics. Yellowstone Park officials still wanted to allow Clover-Mist to run its natural course in their part of the ecosystem.

While Bates had some sympathy for prescribed natural fire from a biological point of view, "from a suppression perspective," he later said, "the strategy made no sense at all." The blaze was spotting in a widely scattered pattern of "mosaic burns," leaving untouched patches of forest, yet its general movement clearly was toward the hydrographic divide between the park and forest, a sparse ridge line cut by densely wooded creek drainages. Bates recognized that letting the fire burn inside the park meant that eventually it would move out of the park "in a big way." It seemed "highly counterproductive," the firefighter understatedly noted, "to fight a fire on one side of the ridge while letting it burn on the other."

But he had his orders. It wasn't his job to settle policy differences between the park and the forest. Bates routed a call to BIFC for five crews (100 firefighters) and four helicopters, and decided to set up a spike camp at Frost Lake, at the boundary line where the slopover had occurred and close to the place the vice-president had been camping a few days earlier. He established his Incident Command Post in a ski lodge near Sleeping Giant Mountain.

"Minimum impact" was to be the byword of the suppression effort in the wilderness area. Wherever possible, natural features such as open meadows, water bodies, and cliffs would be used as firebreaks. Bates' Planning Section chief filed an Incident Command System 202 Form, stating the "Incident Objectives." As always is the case, the first objective was to "provide for firefighter safety." The second objective was to "keep Clover-Mist fire within Yellowstone National Park boundaries." The third objective was to "contain and confine all boundary fires on national forest lands to less than

1,000 acres." Finally, the 202 form noted they were operating in grizzly bear territory. Special precautions would have to be taken in handling food and garbage, and a twenty-four-hour security patrol was established at the spike camp.

On the morning of July 25, Bates flew over the Clover-Mist fire with his Operations Section chief. The fire was active on the boundary line, and Bates was having trouble getting his crews into the high country spike camp by Frost Lake. The helicopters he had ordered had not arrived and he needed them to ferry in crews and equipment: another resource problem. His attempt to "borrow" a helicopter from the Mink fire had not been successful and the lack of helicopters was looming as a major difficulty. His entire suppression effort seemed to depend on them.

Bates also was having problems quelling the murmurings of dissent in his overhead team. The teams, as noted earlier, are the brain centers of the suppression effort, consisting of thirteen to fifteen key personnel responsible for the planning, logistics, supply, financing, and execution of the firefight. Crews and equipment are rotated in and out as needed, but the overhead team stays together. Bates' team strongly objected to allowing Clover-Mist to burn inside the park.

"Day planning meeting developed the concerns over National Park Service management philosophy versus U.S. Forest Service," Bates laconically noted in his diary of the Clover-Mist fire. "This could have a bearing on future plans."

Bates also would express uncertainty about the role of Area Command. "Acknowledged Area Command," he wrote in an early diary entry, "but still unsure of them. They have made no contacts yet." Within two days of his arrival he would fly to West Yellowstone for a meeting with Area Command, yet the uncertainty would persist. An official review panel for the Clover-Mist fire would later conclude that the "role of Area Command was not clearly defined."

Area Command, however, was only part of the problem. Bates and the other incident commanders were serving multiple masters. Bates, the Clover-Mist review would note, "received direction independently from the Park and the For-

78   est." Technically, he knew he was supposed to be working for Area Command, but in reality he "worked for the Park and Forest." Incident commanders "were confused by the lack of unified direction and coordinated strategies," the review noted. "Bates was not clear about Area Command's authority and received very little direction."

Bates' confusion should be understandable. He had just flown into a zone crowded with competing powers. By July 25, there were *thirteen* generals in and around the Yellowstone Nation, all of them hammering for attention, resources, and clear marching orders: Bob Barbee, besieged at Mammoth, had triggered the establishment of an Area Command unit with two commanders, one from the Forest Service and one from the Park Service; Dan Sholly was serving as incident commander for the Fan fire in the north and desperately attempting to keep tabs on all the ongoing operations in his domain; in addition to Bates on Clover-Mist, there were incident commanders on the Mink, Snake River, and North Fork fires; and Barbee's counterparts, the supervisors of five of the national forests surrounding Yellowstone—the Targhee, Bridger-Teton, Shoshone, Gallatin, and Custer forests—all were maneuvering for best position and resources.

Residents of the gateway communities were furious about the spreading fires and confusion, and hundreds—soon to be thousands—of firefighters were pouring into the area. Beyond the immediate confines of Yellowstone, the national media, alerted by the regional coverage of the Mink fire and Dave Poncin's evacuation of Grant Village, were heading to the area. So were Interior Secretary Donald Hodel and Wyoming Senator Malcolm Wallop. In Washington, National Park Service Director William Penn Mott had released a statement strongly defending the natural fire policy.

Others were on the way too. Area Command had summoned the nation's top fire-behavior experts, skilled and dedicated professionals armed with the latest in high-technology equipment and sophisticated computer models. They too would play a critical role in the events of late July, August, and September.

In the firefighting profession, fire-behavior analysts are often known as "fire gods" because of their forecasting abilities and power to recommend, for example, the torching off of a couple of thousand acres in order to prevent a much larger fire. Men such as Curt Bates, Dave Poncin, and Dan Sholly also were sometimes called fire gods, a term mixed with the derision and respect the personnel on the fire line feel for the people running the show. Ultimately, however, all the fire gods would fail, and it was in their failure that the collision of the Old West mystique of the capable man—the ethos of Hemingway and frontier historian Frederick Jackson Turner—with the bureaucratized New West of ecosystem models and natural regulation would become most apparent.

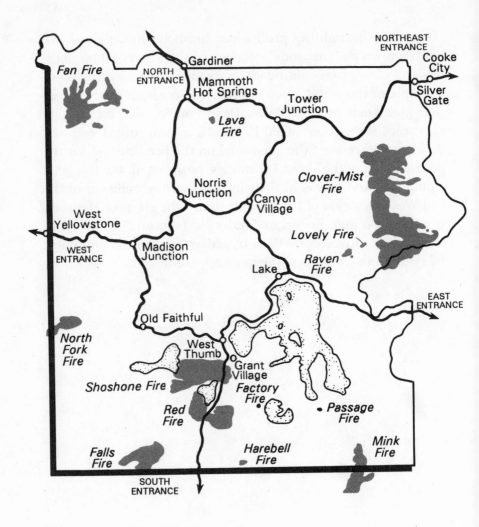

YELLOWSTONE FIRES

July 25

Shoshone fire pushes toward Grant Village;
North Fork fire enters park; Clover and Mist fires link;
Mink fire enters park.

© Linda Marston, 1993

BOOK TWO

Fire Gods

Who was I, a mere park superintendent, to question the wisdom of "the father of fire behavior"?

—ROBERT BARBEE

VIII

Grant Village

Alarmed by the spot fires over the South Entrance Road, Dave Poncin had decided to make a stand at the entrance road a quarter-mile away from Grant Village. That was Plan A. If the fire beat them there, they would go to Plan B, falling back to the loop drive circling the development.

Poncin split his 500 troops into two divisions. They were on the move before the dawn of July 25. The First Division would work a section of the highway southwest of Grant Village, clearing a handline down to mineral soil on the west side of the road and setting a burnout to rob the Shoshone blaze of fuel. The Second Division would conduct a similar operation to the northwest.

Poncin accompanied the First Division. "It seemed they were most likely to have a direct encounter and I wanted a firsthand look at the fire I had been hearing so much about," Poncin recalled. "The crews completed the 6-mile line by noon and soon after the squad leaders were moving along its edge with drip torches, setting the burnout." As the wind began to pick up, two DC-6 air tankers were called in to lay down a slurry drop of retardant, thick pink nitrogen fertilizer.

The fire god was feeling pretty good. "Our defenses looked solid," he said. They had the highway, the handline, a slowly progressing burnout, and the slurry drop in the trees adjacent to the burnout. "The plan was to keep the fire out of the canopy and on the ground, so our crews wouldn't be faced with a high-intensity crown fire when and if the Shoshone fire hit the road."

A curtain of smoke fell across them, behind it the ominous rumble of a gigantic force. "We could feel the heat on our faces. Suddenly an erratic gust—a fire-behavior analyst on the road clocked it at 25 miles per hour—blew a long finger of flame through the treetops. Then there were fingers of flame everywhere, flashing from the smoke." The fire was on a run, crowning out, lofting right over Poncin's defenses. The incident commander stood on the road, directing the attack on the spot fires as the rain of fire came down and blew on past, into the forest behind him, moving for Grant Village.

They chased the spots, but there was too much fire and not enough firefighters. "We were outnumbered," Poncin says. "We were losing the line. I looked down the highway: the summer day had been turned into an orange night of spot fires and smoke. 'So much for plan A,' I thought."

In Dave Poncin's first direct encounter with the Yellowstone fires, all the briefings and fire-behavior reports were quickly eclipsed by the evidence in front of his eyes. "It was a tough fire, erratic, hot and dirty, with heavy fuel and strong winds," Poncin says. "My team had fought fire all over America, but I couldn't remember the last time we had seen such unusual behavior. We should have been able to keep the fire on the ground; instead it ran right through the canopy and jumped clear over us."

Poncin and his crews fell back to Grant Village. It was time for Plan B.

Shortly after 5 P.M., the Shoshone fire hit the loop road and crossed into Grant Village. Poncin's firefighters had been at work since dawn. Hose had been laid, debris scattered from

buildings, and personnel assigned to each structure. The fire
swept in and the water and foam trucks went to work, spray-
ing the buildings. Spot fires appeared and were extinguished.
Helicopters lifted great buckets of water from Yellowstone
Lake and dumped them on hot spots inside the development.
A DC-6 careened over the village, laying down great clouds of
pink retardant.

At the base camp by the lake, three fire trucks stood
guard as the cooks and camp personnel prepared dinner for
500. Planes and helicopters chased overhead and the rumble
of the fire was all around them. "One of the cooks," Poncin
remembers, "was carrying a box of food to the big mess tent
when he felt a sudden hot gust of wind. He looked up. The fire
was in the treetops a hundred feet away and coming right at
them. Time to head for the lake!"

The forty camp personnel dashed into the shallow waters
of Yellowstone Lake as firefighters scrambled to turn the high-
pressure water cannons on the mass of fire in the canopy. The
engine crews battled the oncoming blaze, blowing it out of the
trees and forcing it north around the camp, where it devoured
a campground restroom, ran down to the lake, and stopped.
The cooks and their camp crews trudged out of the water.
Within an hour, amid the smoking ruins of the forest, dinner
was served.

Ironically, the *destruction* of Grant Village would have
been viewed as a covert victory for environmentalists, who
oppose the lakeside development. The Greater Yellowstone
Coalition considers Grant Village a hideous eyesore and a
threat to grizzly bear habitat and local streams. At the end of
August, Earth First! cofounder Howie Wolke was quoted in
the *Jackson Hole News* as saying that it was a "real shame"
that Grant Village had not gone up in flames. Many Park Service employees had privately echoed the sentiment.

But by the evening of July 25 it was clear that Grant Village had survived the initial assault of the Shoshone fire. An
inversion had formed over the area, the cold air mass pressing
down over the Snake River Complex and preventing nighttime
flare-ups. Yet large islands of dry unburned fuel remained

86   inside the development and that had Dave Poncin worried. He decided that come morning they would burn out the remaining fuels. It was a dicey operation. A high wind at the wrong moment could kick the fire right back on top of them.

As dawn lightened the smoke-filled camp on July 26, Poncin put his fuel-removal strategy into motion. He needed to move fast, before the winds returned. His crews hustled through the morning, clearing brush, building and securing lines. Then he positioned them in guard posts throughout the development. A helicopter swept in and began firing an aerial-ignition gun called a "ping-pong-ball dispenser" into the unburned areas. "Ping-pong balls" are small spheres filled with potassium permanganate which are punctured by a needle that inserts a few drops of ethyl glycol as the helicopter nears the target, creating a tiny napalm grenade. The dispenser drops the ball out, it sails into the woods, ignites, and burns for about ten seconds.

On the ground, Poncin's crews were moving along the lines, setting fires with drip torches while the woods beyond them were bombed with ping-pong balls. Poncin also had bought "a little insurance" by bringing in a pickup truck mounted with a flamethrower called a "terra-torch." On July 26 the fuel-reduction plan went smoothly, and the following day was spent "zapping out" the few remaining islands of fuel along the loop road with a hundred-foot-long tongue of fire from the terra-torch.

Plan B had worked. Grant Village had been saved. The area was fireproofed and the Shoshone blaze had been diverted around the development to the north and east. The smaller tourist outpost of West Thumb, north along the lake, was still under threat, but overall the situation was looking good, and park officials were delighted. It was a significant victory in what had been a very bad week.

"Poncin's victory was very important to us at the time," Barbee recalls. "It contributed to the sense that maybe the worst was over." To add to the growing optimism at Mammoth, a light rain had begun to fall. At Old Faithful and West Yellowstone, the fire gods were assembling: Area Command

was taking over fire coordination; the national fire-behavior 87
experts had arrived; and a political fire god, Interior Secretary
Donald Hodel, was flying in to express support for Yellowstone's natural fire policy. The Yellowstone Nation savored a
delicious moment of resurgent confidence. Unfortunately, a
moment was all it would turn out to be.

IX

Old Faithful

Once again—hard on the heels of the changes in fire behavior and the increased suppression effort—the dynamics of the Yellowstone fires were abruptly shifting. The story itself was moving from a regional event to a national one. By the time Interior Secretary Donald Hodel arrived at the Old Faithful Inn on July 27, two new players, Washington politicians and the national media, had been drawn into the story, and were about to drive it to a new height of confusion. On the political front, Wyoming senators Malcolm Wallop and Alan Simpson—both long involved with Yellowstone issues—had been receiving complaints about the fires from their constituents since June, complaints they had been relaying to Bob Barbee. When smoke from the Mink fire enveloped Cody, Malcolm Wallop flew out for a firsthand look at the suppression activities. The smoke was so thick at Cody's airport that Wallop's plane had to land on instrument navigation. Wallop was assured that everything possible was being done to fight the fires.

On July 25, the story had gone national. "The first media blitz came with the Shoshone fire's threat to Grant Village and the movement of North Fork into the park," recalled Yellowstone's director of Public Affairs, Joan Anzelmo. "Media inquiries went from seven per day to more than thirty on the afternoon of July 23. A local free-lancer had called in North Fork's threat to Old Faithful to the *Denver Post*, the wires picked it up, the networks noticed, and suddenly Washington started paying closer attention."

NBC and the *New York Times* were the first major media to run the story. NBC reported on July 25 that "wildfires continue to burn at Yellowstone National Park," and featured vivid shots of flaming trees, helicopter water drops, and firefighters. The *New York Times* ran an Associated Press story quoting Forest Service spokesman Brian Fox. "Basically," Fox said, "we've decided to allow nature to take its course." At Yellowstone's Public Affairs Office, media requests for information surged. Anzelmo recruited Park Service volunteers to help with the deluge.

In Washington, a subtle shift in the political dynamic had occurred. National press attention had upped the political ante. The Park Service and its parent organization, the Department of the Interior, were coming under increasing pressure. For weeks, members of the Wyoming and Montana congressional delegations had been passing along to Interior the simple message they were receiving from their constituents: put the damn fires out. This was the message reporters were now starting to hear.

Interior Secretary Donald Hodel was in a tricky position. He didn't want to offend the influential Wyoming senators, yet his own Park Service director, William Penn Mott, was sending out an entirely different and equally passionate message: let the fires burn. Mott viewed the Yellowstone fires as an excellent opportunity to educate the public about the positive role of fire in the ecosystem. So while Wallop and Simpson were responding to the passions of their constituents in pressing for greater fire suppression, Mott was responding to the passions of *his* constituents—the National Park Service and environmentalists—in pressing the case for natural fire. Hodel's task was to placate the Western delegations and support the firefighting effort, yet not cut the ground out from under Mott and the natural fire program which, after all, was official Park Service policy. Hodel decided to fly out to Yellowstone for some on-the-spot damage control.

Barbee met him at the West Yellowstone airstrip. Hodel immediately asked Barbee what he, Hodel, ought to tell the press. Barbee was acutely aware of the various pressures on Hodel, yet the mixed signals over natural fire and fire suppression were making him uneasy. They needed, somehow, to get an unambiguous message out. Yet they also needed to protect the policy. "I told Hodel that we should announce that we're not allowing the fires to burn," Barbee recalled. "We had to tell people that fire has a natural role to play in the wilderness, but at the same time assure them that we were stopping the natural fire program because it seemed to be developing into an extraordinary year."

Like Donald Hodel, Bob Barbee had more than one constituency to play to. He wanted the primary message to be that the fires were being actively fought. He needed to boost the sagging morale of his own troops and reassure the local communities and park concessionaires that Yellowstone wasn't going up in smoke. But Barbee couldn't ignore the wishes of Mott, who wanted the natural fire program defended. And for all his doubts about the immediate political efficacy of natural fire, Barbee still believed in the mission: fire was an integral part of the ecosystem. The park superintendent was maneuvering to rescue Yellowstone from the present crisis *and* preserve the natural fire policy for the future.

The sun broke through the clouds as Donald Hodel, his button-down shirt open at the neck and his sleeves rolled up, stepped before the cameras and began a defense of Yellowstone Park and the natural fire policy. He had just completed a helicopter tour of the park. He was, he said, "impressed by how much of Yellowstone is not burned." Hodel spoke to about eighty reporters assembled in the parking lot beside the Old Faithful Inn. The fires were being fought, he said. But, he added, "we aren't going to waste our resources where fires

92   aren't doing harm to the park. There is a long-term beneficial effect from fire." In the distance, a faint rumble from the North Fork fire could be heard. The blaze was expanding. A helicopter cut overhead.

Hodel pointed out that fire clears away diseased trees, replenishes the soil with nutrients, creates new habitat for wildlife, and regenerates flora. Fire was an essential part of the forest's life cycle. "This is the historic pattern of Yellowstone National Park," he said. "What we enjoy here today is a result of that kind of historic happening." He saw no reason to change the natural fire policy, but indicated there would be a "clarification" of it on a "national basis." In response to a reporter's question, Hodel said the park did not let the fires "get out of hand." The fires were not out of control; they were being "effectively managed." Minimum-impact firefighting was the proper procedure in Yellowstone. "These fires sometimes will leap a mile ahead of themselves. If you were to try to put in a fire line wide enough to prevent that happening, you'd have to get in there and bulldoze a mile or more wide. Let me tell you, you *would* see that for the next 300 years!"

A local rancher stepped forward to present a petition calling for more aggressive firefighting. It was signed by 300 high-country residents. They were tired of breathing smoke, the rancher said, and were worried by such intense fire behavior so early in the season. "If a fire started the first of October," the rancher told Hodel and the press, "that would be a different story. But when a fire starts now, the thing could burn to the fifteenth of October. That's taking a lot of chances."

Hodel accepted the petition. The press conference was over. The reporters scattered. Bob Barbee was delighted by the secretary's performance. He had supported the park, supported the policy, and given the press a good lecture on fire ecology and minimum-impact fire suppression. He had danced in the direction of placating the local communities and the Wyoming senators by indicating a "clarification" of the fire policy, casting it as a national reevaluation, while adding that the fires were good "as long as it doesn't come to the point where too much of the park, too vast an area, is adversely affected." It was a masterful performance. Hodel had given everybody a little something and closed out no options.

"Hodel was being bludgeoned by his ideological soulmates in the political world," Barbee recalled later. "Many of them simply wanted the fires out—period. But Hodel stood by his Park Service troops. For that, he earned our enduring gratitude."

But while the press conference was politically deft, essentially it was another barrage of mixed signals: fire is bad and we are going to put it out; fire is good and we are going to let it burn. The media coverage would reflect this. Some, such as NBC (which with the exception of CNN devoted the most news minutes to the story throughout the summer), would lead its report of the Hodel visit by announcing that "firefighters reverse policy and use aggressive tactics at Yellowstone," creating the impression that Hodel had ordered a policy reversal. He had not. Yellowstone had shifted to the suppression options within the natural fire policy almost a week earlier, but as we've seen, that move was lost in the confusion over fire-suppression terminology (such as "confinement") and mixed signals from the national forests.

Others in the media emphasized Hodel's buttressing of Yellowstone and the fire policy: "Hodel Supports Yellowstone's Natural Burn Policy on Fires," announced the headline in the *Billings Gazette;* "Hodel Visits Yellowstone, Downplays Fires" ran the story in both the *Boston Globe* and California's *San Jose Mercury News.*

"In our minds," Barbee says, "we had taken major steps by moving to suppress some fires with our own forces, and calling in incident management teams and Area Command. Yet all the public saw was pictures of burning forests."

And neither the park nor the press commented on the fact that there was a confluence of Yellowstone's natural inclination to let fires burn with the main tactical option open to firefighters at the time, the tactic of confinement: letting a blaze go until it hits a natural barrier.

In retrospect, perhaps the Hodel press conference called

for a bit of political theater, a performance of absolute moral and lexical clarity: no dancing in the direction of natural fire and long-term beneficial ecological effects, but instead only heroic firefighters in an old-fashioned battle against the elements. But this, after all, was ineluctably the New West, with its slightly smug air of scientific certainty, and the players were tied to the mission, and the mission was tangled in a web of politics. Now, however, the media wanted some answers.

In the weeks ahead, Yellowstone officials would come to see the mixed messages of the press not as a reflection of the tumult inside the park but as evidence of a new enemy assault. The fires were bad enough, but now it seemed to park officials that the media were out to get them. And while many Yellowstone officials would continue to dispute the contention that the fires were an environmental crisis, few in August and September would deny that they had slipped into a full-blown public-relations calamity.

X

West Yellowstone

The maximum fire gods were gathering. By the end of July, the Area Command group had set up operations in the gateway community of West Yellowstone and was waiting for the report and forecast from the fire-behavior analysts. In addition to Area Command and the analysts, another unit, virtually unheard of during the summer of fire, had just visited the scene. To fire cognoscenti they were known as "Big Mac"—technically a "Multi-Agency Coordination group"—shadowy fire gods from some firefighting Olympus beyond the ken of mere mortals.

"Talk about your fire gods!" remembers Bob Barbee. "This was where real power resided. Big Mac set regional, and sometimes national, firefighting priorities." While Yellowstone Park and incident commanders had to go to Area Command for resources, Area Command had to go to Big Mac. Forest fires were now burning in Montana, Wyoming, Utah, Idaho, South Dakota, California, Washington, Wisconsin, and Alaska. Big Mac was a kind of triage unit—it decided where the resources would go, which forests would be saved, and which would burn.

Big Mac's visit to West Yellowstone to discuss Area Command's scope and strategy was an encouraging sign that Yellowstone Park was the top national priority. Another sign of Yellowstone's importance was the appearance of a legendary figure to lead the fire-behavior analysts. His name was Dick Rothermel; the fifty-nine-year-old scientist, based at the U.S. Forest Service Intermountain Fire Laboratory in Missoula, Montana, was known as "the father of fire behavior." He had educated a generation of fire-behavior analysts, written widely on the subject, pioneered the use of wind tunnels in studying fire, and developed the key model for computing fire spread.

Rothermel was joined at West Yellowstone by five other top fire-behavior analysts, including Yellowstone's Don Despain. They were given an urgent task by Area Command: predict the likely scenarios in each of the major fire complexes. The report needed to be ready in seventy-two hours, on August 2, when Jack Troyer, the team leader of the Greater Yellowstone Coordinating Committee, was bringing together a major meeting of Area Command, park superintendents, forest supervisors, incident commanders, and other key personnel.

At Mammoth Hot Springs, Bob Barbee and Dan Sholly viewed the August 2 meeting with some trepidation. They were ambivalent about Area Command, which was poised to usurp their power. Although cooperating with Area Command, Sholly continued to closely monitor all the fires and run what in effect was a second firefighting command post out of Mammoth Hot Springs. Barbee and Sholly were team players, and the mission now called for interagency unity, but they continued to believe that they knew what was best for Yellowstone Park. Still, they would try to abide by Area Command decisions.

"I was no stranger to fire," Barbee says, "but I wasn't a fire god on the level of Dick Rothermel either. I was prepared to trust and support the experts, provided there was no gross collision with park policy." Trusting the experts, as it turned out, was a grave mistake. The fire gods were heading for a very big fall.

For three days and nights the fire-behavior team worked in Area Command's headquarters in West Yellowstone, poring over Despain's detailed maps of the Yellowstone fuel complexes—exhaustive delineations of tree and undergrowth types, growth stages, and locations—as well as topographic maps, decades of precipitation records, infrared fire surveys, historical fire records, field reports, and fire-spread calculations. When all the fire gods assembled at the Area Command headquarters at the West Yellowstone airfield on August 2, the fire-behavior experts brought good news: the worst was over. It was going to rain. And while the fire behavior team's directive was only to make projections, not recommendations on strategy, the conclusion was obvious: the fires should be allowed to burn.

Curiously, written records of the important meeting seem to have vanished. According to several sources present at the gathering, however, the fire-behavior team's main contention was that Yellowstone precipitation records indicated that rain in August was almost certain. Furthermore, the movement of the fires, due to the prevailing southwest winds, was along a predictable northeastward path that did not pose an immediate or short-term threat to lives or structures. But the joker in the deck, the fire-behavior team noted and Don Despain confirmed in later interviews, was the wind. Based on the weather records and the data at hand, the team predicted that at most there might be one major wind event—a "wind event" being defined as winds over 40 miles per hour for a duration of over two hours.

Turning to specific fires, Rothermel and Despain concluded that the North Fork fire stood little chance of hitting Old Faithful to the east, or West Yellowstone, more than 12 miles north across the Madison Plateau. The prevailing winds would carry the fire northeasterly, missing Old Faithful. As to any threat to West Yellowstone and the popular road between it and Madison Junction in the park, the northern reaches of the Madison Plateau were scattered with large areas of young lodgepole pine and meadows, through which a fire would have trouble traveling. The young fuels were too green and the fuel on the forest floor too sparse to effectively carry the blaze. North Fork would not hit the Madison road.

According to participants in the meeting, the North Fork incident commander, Larry Caplinger, strongly disagreed with the analysis. It seemed to him that the drought was causing the young fuels to carry fire the way older fuels usually did. Also, the winds were much too unpredictable. Strange things were happening out there. Over recent days the wind had driven the North Fork fire another 4 miles north, past the Little Firehole Meadows and toward a place called Buffalo Meadows; the south and east flanks, near Old Faithful, continued to hold. A huge convection column towering at the horizon was making the residents of West Yellowstone nervous.

Dave Poncin, incident commander of the Snake River Complex, also had been alarmed by the unusual fire behavior. "On August 2, while we were meeting at West Yellowstone," Poncin later recalled, "southwest winds gusting up to 30 miles per hour were driving the Shoshone fire 3 miles east from Delusion Lake to Eagle Bay on Yellowstone Lake." It was the third major run of the fire since Poncin's successful defense of Grant Village on July 25.

South of the Shoshone fire, the other blazes in the Snake River Complex, the Red and Falls fires, also had been active, Poncin noted. The Red fire—now at 16,000 acres—had burned around Factory Hill, over the Witch Creek drainage, and advanced toward the Heart Lake Patrol Cabin. Red's southern cousin, the Falls fire, had broken through a control line on its eastern flank on July 29. That day a new fire, the Continental, was reported on the South Arm of Yellowstone Lake. With resources low and safety a major concern, the park allowed it to burn for two days with aerial monitoring, then attacked it, due to mounting political pressure against letting a new fire burn. Extreme fire behavior and lack of support resources caused the park crews to be pulled off almost immediately. "In fact," said one firefighter at the scene, "we had to run for our lives."

Poncin also seconded Larry Caplinger's observations of the "fine fuels"—meadow grasses, undergrowth, and duff—and young fuels. He said that the fuels seemed to have "significantly dried up" during his time in the park. What had once served as a barrier to the spread of extreme fire now, under windy conditions, would have exactly the opposite effect.

The fire-behavior team, however, held to its opinion that rain and natural barriers such as Yellowstone Lake would keep the Snake River fires within reasonable limits. Rothermel and Despain also were optimistic about the two major fires in the far north and south, Fan and Mink. In the south, Mink had burned into the park and was nearing Cliff Creek, but light-hand tactics seemed to be working well against it. In the north, the Fan fire had increased to 15,000 acres and was threatening to move out of the park and on to private land owned by a controversial New Age religious group often at odds with Yellowstone Park, the Church Universal and Triumphant. Its leader, Elizabeth Clare Prophet, also known as "Guru Ma," was lining up her followers in a field within sight of the fire, where they would soon engage in marathon, ultimately successful, high-speed chanting to turn back the blaze. (In early August—ironically, given the history of the church's feuds with the park—Barbee would press Area Command to take more aggressive action against the Fan fire.) Rothermel and Despain, however, were confident that the Gallatin Range and precipitation would slow Fan's growth.

Regarding the eastern sector of Yellowstone, where Clover-Mist, now at 73,000 acres, was the park's biggest fire, the fire behavioralists were equally sanguine. Rain, and the great ridges of the Absaroka Range, would contain the fire.

It was not an opinion shared by Curt Bates, incident commander on Clover-Mist. He viewed it as essentially the same line the park had taken since the start of the fires. Anticipating the Area Command let-burn decision and preparing to move on to other fires, Bates had prepared a Contingency Analysis to deal with his anticipation of the fire's spread. In it, Bates recommended burnout operations and retardant drops if the fire passed certain geographic trigger points. Bates and *his* fire-behavior analysts were concerned that the fire would move up the creek drainages, through the mountain passes, and out of the park, threatening Forest Service land, the ranches of the Clarks Fork Valley, and the gateway communities of Silver Gate and Cooke City.

"We recommended specific ridges between the fires and the Yellowstone Park boundary as the trigger points for a more aggressive suppression effort," Bates recalled later. "The strategy, if initiated at the proper time, would have confined the fire inside Yellowstone Park."

Curt Bates had completed his mission: he had contained the small part of the Clover-Mist fire on the Shoshone National Forest, while allowing the major part of the blaze to burn inside Yellowstone National Park. On August 2, Bates demobilized and turned the biggest fire in the park over to a twenty-person strike team under Dan Sholly. Departing the scene, Bates again noted in his diary that "differing philosophies between agencies is frustrating [and] leaving a large fire still uncontained is very frustrating." As it would turn out, Bates' fears and frustrations, as expressed in his diary and in the Contingency Analysis he left behind, were completely justified.

Snake River still burned, Clover-Mist still burned, but Dave Poncin and Curt Bates were on the way out. Poncin, Bates, and Caplinger had made their doubts known. For Poncin and Bates, their jobs—protect Grant Village, contain the portion of Clover-Mist on the national forest—were done now, and their Type I teams were needed elsewhere. Big Mac was calling, and an important gear in the complex machinery of wildland firefighting was kicking in: transitions. Within a day of the August 2 Area Command meeting, both Poncin and Bates would have "transitioned out" of the region. Poncin turned Snake River over to a Type II team led by Incident Commander Dave Fischer—while excellent firefighters, Type II teams usually are deployed on lower-priority, less complex fires. "See you at Madison Junction," Poncin had joked at the end of the August 2 meeting, in what turned out to be a prophetic glimpse of the North Fork fire.

By the end of the fire season, *twenty-six* incident command teams, Type I's and Type II's, would have transitioned through the Yellowstone area, and valuable time would be lost in briefing the officers, bringing them up to date on the fire they had to deal with, familiarizing them with the terrain, and explaining the light-hand procedures so important to Yellowstone Park. And, shortly, both Dave Poncin and Curt Bates would be back, taking command of huge blazes the fire behavioralists had failed to predict.

Can blame for the Yellowstone fires be pinned on the fire-behavior team led by Dick Rothermel and Don Despain? The analysts were in a tremendously difficult and exciting position. By the end of July, it was clear that 1988 was shaping up as a historic fire season. But fire prediction is a notoriously inexact science, one usually scaled to a matter of days, not the one-week, two-week, and longer projections requested by Area Command. These projections created a huge margin for error.

Yet were the fire behavioralists perhaps seeing only what they wanted to see? Much anecdotal evidence conflicted with the main thrust of the report. Although Rothermel and Despain, basing the forecast on past fire-weather history, anticipated only one major wind event for the upcoming weeks, the previous weeks—and even the previous days—had revealed incidents of erratic and gusty winds. Clover-Mist, North Fork, and Shoshone all had periods of erratic wind-driven movement; the new Continental fire had exhibited extreme fire behavior; and Dave Poncin was reporting threatening gusts inside power-line corridors and on the east flanks of all the Snake River fires. Furthermore, longtime residents of the Greater Yellowstone Area had abundant experience of localized "microclimate" winds created by the steep mountain ridges and creek drainages. But relations with the local communities were poisoned by mutual distrust, hampering the flow of information from a potentially valuable source.

The matter of fuels also was problematic. Projections on

102   the North Fork fire were based on the belief that it would
rain and that the fire would slow down when it hit young
fuels. Yet two incident commanders, Larry Caplinger and
Dave Poncin, had reported a considerable drying up of fine
fuels and young fuels. These fuels seemed to be carrying fire
at an accelerated rate. Don Despain himself had witnessed
unusual fire behavior in a growth of young lodgepole pine
near Grant Village.

It is difficult to escape the conclusion that the August 2
report of the country's top fire-behavior experts, after more
than six weeks of fire in the area, continued the bias in favor
of the natural fire policy. Yellowstone Park, in the view of its
scientists, was at a profound ecological crossroads. This view
made a strong impression on Area Command, incident commanders, and Robert Barbee. In the end it was they—not the
fire behavioralists—who would be responsible for the tactical
decisions. And from a tactical firefighting point of view, "con-
fining" the fires—in reality another way of letting them
burn—made sense. Precipitation records did indicate it would
rain. Contrary evidence—winds, fuels, the dissent of some
incident commanders—did not carry the day.

Somewhere, alarm bells should have gone off. What
about the drying of fuels? What about the massive size of the
Clover-Mist fire and Incident Commander Curt Bates' strong
objections to the firefighting policy? What about Bates' call
for more aggressive action? What about the rumors swirling
around Mammoth Hot Springs that Yellowstone Park firefighters—the people, after all, who *did* know the terrain—
were growing increasingly opposed to letting *any* of the fires
burn?

But alarm bells did not go off. In part, this was because
the scientists were running the show. "Who was I," says Barbee, "a mere park superintendent, to question the wisdom of
the 'father of fire behavior,' the maximum fire god himself?"
Barbee accepted the recommendations to let the fires burn,
using limited suppression strategies.

In part, also, the findings went unchallenged because it

was what the park wanted to hear. It would rain soon; the fires would be contained; the damage would be controlled. It followed, then, that the natural fire policy would survive, which also meant park officials would survive.

Somewhere back behind the clouds, the great goddess Fire herself must have been chuckling as she filled her cavernous cheeks with wind. For it would only be a matter of days before the hubristic forecasts of the Yellowstone fire gods would be blown to pieces.

North Fork

Eight days after transitioning off the fires, Dave Poncin and his Type I team were back at West Yellowstone being briefed by Area Command. A light rain on August 3 had reinforced the fire gods' faith in the long-term weather records and Area Command had begun to demobilize and downgrade some of the fires, "releasing" the resources for firefights elsewhere. By August 6, however, a thunderstorm bringing 60 mile-per-hour gusts had blown in, causing a burnout operation at the north end of the North Fork fire to run out of control.

The "one major wind event" of the Rothermel–Despain forecast had arrived. Unfortunately, there would be many more. By August 7 the fire from the burnout operation, and the North Fork fire itself, was bearing down on the key park crossroads at Madison Junction. Another gusty thunderstorm was forecast for August 10. And Area Command itself was being buffeted by organizational and psychological transitions. As the fires appeared to decrease in intensity during the first days of August, Area Command released much of the public-information and support staff. The Big Mac group flew in and approved continuing reductions. But the thunderstorm of August 6 and the news of another approaching storm front were by August 10 causing yet another reevaluation, and Area Command began to reverse gears, building up again. Meanwhile, transitions at the top of the Area Command structure also were underway. New Forest Service and Park Service co-commanders were installed.

Area Command's new assignment for Incident Commander Poncin was the North Fork fire. Poncin had returned to an arena of extreme organizational flux at a time when continuity ought to have been the name of the game. Despite the efforts made at interagency cooperation, the situation at the command level was chaotic. The chaos was shielded from the public, but Yellowstone Park viewed it with increasing dismay. Chief Ranger Sholly tried to impose control and continuity on the situation by delegating "line officers" to Area Command and the incident command posts, instructing his rangers to keep a sharp eye out for violations of the firefighting guidelines, and continuing to serve as incident commander on several fires himself. But his boss's sense of embattlement was growing. Barbee was keenly aware of the rising confusion and resisted surrendering tactical control of the Yellowstone Nation to a constantly shifting cast of ambitious officers, most of them from the rival Forest Service.

A microcosm of this institutional power struggle can be found in the subtle differences in the briefings Dave Poncin received as he twice transitioned on to the Yellowstone fires. In the first briefing, on July 23, as he came in to take over the Snake River fires, Poncin met with Barbee and Sholly at Mammoth Hot Springs. After a discussion about what exactly was meant by "light hand on the land" tactics, Poncin was given his marching orders—essentially, protect Grant Village—and left to do his job. In bureaucratic terms, the "delegation of authority" was fairly straightforward: from Barbee to Poncin. Eighteen days later, as he came in to take over the North Fork fire, things were not so simple.

Now the Forest Service, via Area Command, was playing

a greater role in the overall effort. The gateway community of West Yellowstone was threatened by North Fork and Old Faithful was not in the clear. Taking over North Fork, Poncin received his briefing not at Mammoth but at Area Command headquarters in West Yellowstone. Although Mammoth remained an important player, the balance of power had shifted away from it.

Poncin was now given *two* delegations of authority, one from Area Command (with the assent of Yellowstone Park) and one from Targhee National Forest Supervisor John Burns, the latter with good reason considering his property still in danger. The two delegations of authority put Poncin on notice that in practical terms he had three satrapies to deal with: Area Command, Targhee National Forest, and Yellowstone National Park.

As was his custom, the incident commander drove to the fire zone as soon as the briefings were completed. "The funny thing was how calm it all seemed," Poncin remembers of his drive from West Yellowstone to Madison Junction. "It was beautiful."

A warm breeze had cleared away the low smoke and beside the road was the sparkling water and grassy meadows of the Madison River. Moose, elk, and bison grazed by the riverbank. A fisherman waded in the stream. Bob Barbee had succeeded in preserving an aura of normality. It all seemed so pastoral, if you ignored the huge convection column leaning up from the south. That was North Fork, deep in the Madison Plateau.

By August 11, Poncin had received the go-ahead for his strategy to fight North Fork. Taking advantage of the few meadows on the Madison Plateau, the roads from Old Faithful and Madison Junction, and the lines Larry Caplinger had put in on parts of the south and east flanks of the fire, Poncin would build what would turn out to be a fire line of epic proportions. It would begin near the original ignition point of the fire on the Targhee National Forest, dip south, then drive 25 miles north and east across the plateau, flanking the fire, tying in to Caplinger's lines, and meeting the road north of Old

Faithful midway at the Midway Geyser Basin. From there it would parallel the road to a pretty canyon spot called Cascades of the Firehole. The idea was to keep North Fork away from Old Faithful and on the Madison Plateau. After the fire escaped control efforts, Poncin's crews would march the line 15 miles east across Yellowstone's Central Plateau. Winding through thick forests, down one plateau and up another, Poncin's fire line eventually would stretch over 100 miles.

There was a second, equally critical part of Poncin's strategy too: keep North Fork south of the road between Madison Junction and West Yellowstone. "If the fire continued burning north, as we expected it to, we would use aerial ignition devices to burn out the woods on top of the cliffs of the Madison Canyon, south of the Madison River," Poncin recalls. "The burned-out woods, together with the high cliffs, the Madison River, and the road, formed a firebreak a quarter-mile wide. Surely that would stop North Fork!"

North Fork continued to run more north than east, avoiding Old Faithful. On August 12, light showers slowed North Fork's advance and gave the firefighters a little breathing room. "There was now absolutely no question in my mind that the fire would hit Madison Junction," Poncin says. "I still hoped to stop it there, at the Madison Canyon cliffs. We would use helitorches"—flamethrowers suspended from helicopters— "and ping-pong balls to ignite the burnout fires along the clifftops that ran in a curve south of the Madison River and west of the Firehole River. The cliffs narrowed into a wedge at the rivers' confluence. North Fork was heading right into that wedge, right for the campground at Madison Junction."

The burnout operation commenced on the afternoon of August 14. The Madison campground was evacuated and the helitorch team swept up the Firehole River canyon, igniting the forest. "The dry timber roared to life," Poncin says, "and a line of black smoke rose from the cliffs." Soon there were embers and ash lofting up in the air and floating on the warm breeze, drifting out over the meadows, over the river, over the road. Watching from the highway, Dave Poncin walked over to a small spot fire sputtering in the grass and stomped it out. "It was not a good sign," he said. He posted a crew to handle the spot fire.

The morning of August 15 brought another "wind event," a dry cold front, gusting up to 40 miles per hour. By noon, the giant head of the North Fork fire—itself now a mixture of the original human-caused blaze and various burnout operations—was attacking along a 10-mile perimeter. "The northeast corner of the fire had jumped the [August 14] burnout and was down in the Firehole Canyon, which was acting like a wind tunnel and funneling the fire north toward Madison Junction," Poncin remembers. "At the same time, the more northerly sector of the head had arrived at the rim of the Madison Canyon." In the twelve days since the fire gods' optimistic assessment at West Yellowstone, North Fork had moved 15 miles across the Madison Plateau and arrived where it was never supposed to be.

Poncin had driven to a bridge by the Madison campground. His radio crackled with electronic traffic as field observers and crew bosses tracked North Fork's run. By 2 P.M. the fire was tumbling off the Madison cliffs and down into the meadow where in 1870 the Washburn Expedition had spent half the night excitedly discussing the idea of creating the world's first national park. "It was awesome," Poncin says. "Fingers of fire were rolling down from the bluffs and I saw the wind lift the fire up, lofting it up, sending great banners of flame flashing out from the smoke. I'd never seen anything quite like it: the fire was blowing out *horizontally*, right off the cliff face! It was a spectacular sight. The very heavens seemed afire."

The smoke descended on the bridge, hiding the true distance of the fire. "The wind set up howling with a fury," Poncin says, pelting the area with spot fires. He grabbed a shovel and began to throw dirt on the spot fires—a moment not without an element of black comedy, considering the inferno rolling off the cliffs a short distance away. Then the incident commander wiped his brow and pulled the radio from his belt. "Operations," he yelled above the roar, "this is

110   the IC. I'm down here on that spot by the bridge and could sure use a hand. Send me an engine!"

Operations Chief Bob Meuchel came on the radio and told Poncin not to bother fighting the spots—they already had lost all their objectives. "The fire had crossed the roads and uncontrolled spots were spreading into large fires and running northeast up the Gibbon Canyon," Poncin remembers. "It was pretty bad."

"Not a good day," Poncin would later note in his diary. "Got our collective butts kicked. A lesson in humility."

Not a good day at all. Poncin said nothing at the time, but he knew the war was lost. Up at Mammoth Hot Springs, Bob Barbee thought the same thing when he heard that North Fork had roared off the Madison cliffs. "No act of man would stop the North Fork fire now," Barbee later said.

Madison Junction signaled a dangerous new phase of the Yellowstone fires, and not at North Fork alone. It was part of Yellowstone Park's staggeringly bad luck that several other fires were independently yet simultaneously undergoing the same personality change as North Fork. The fires were growing more unruly and more unpredictable. The general fire pattern was now moving off the open plateau country, where the fronts burned in fairly even perimeters, and into a jagged land of canyons and creek drainages. There was more spruce, which had a greater tendency to spot. The winds whipped and switched and rolled in treacherous ways. The duff burned out from below boulders and logs on steep slopes and sent them crashing down on firefighters. It was a new game now, even deadlier than before. Here was where the Yellowstone fires would pass into history.

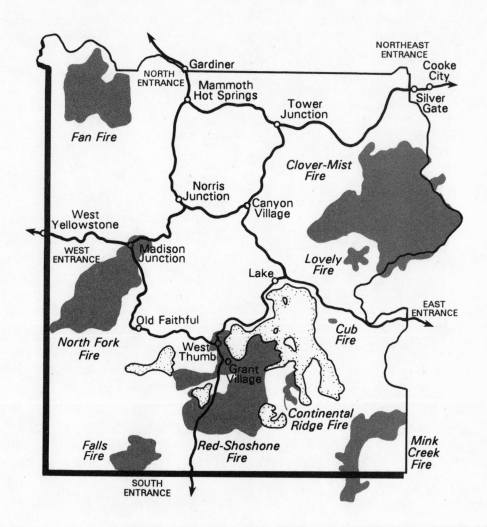

YELLOWSTONE FIRES
August 16

All fires consolidate and expand; North Fork fire hits
roads and moves past Madison Junction.

© Linda Marston, 1993

BOOK THREE

Black Saturday

*SAFE PRACTICES UNDER BLOWUP
CONDITIONS
Consider possibility of retreating into burn.*

—*FIRELINE HANDBOOK*

XII

Into the Black

Mammoth Hot Springs
3 A.M.–4 A.M.

By mid-August, the wind had taken complete control of the Yellowstone fires. A new blaze, started at an outfitter's camp on Hellroaring Creek, had been pushed 4 miles in just four days by intermittent 30-mile-per-hour winds. On August 19, fire scouts reported a troubling sign: whirlwinds were lifting "large material" from the Hellroaring fire and hurling it hundreds of yards. BIFC weather satellites confirmed the bad news: another thunderstorm was on the way. Extreme winds were expected. A red flag warning, the highest state of alert, had been posted for Saturday, August 20.

Waiting in the night at the gates of fire—at Mammoth and West Yellowstone, Old Faithful and Madison Junction, at command posts and spike camps scattered across the great wilderness sanctuary—were 2,800 firefighters, 240 engines, and 26 planes and helicopters. No one knew how big the wind of August 20 would be, but they all knew it was out there, building up beneath a high-pressure front over the Great Basin. The fire perimeters in Yellowstone seemed to know it too—observers recall that the perimeters seemed to flicker hungrily in the night. Soon an avalanche of air would come crashing down on them. It had been only five weeks since Dan Sholly deployed shelters at Calfee Meadow, but things had utterly changed. Vast cities of flame, over 300,000 acres, were now spread across the blackness of the Yellowstone night.

In its present circumstances, Yellowstone scarcely could be said to be a "wilderness" in the sense outlined by the Leopold Report and painstakingly worked into national policy by a generation of environmentalists—that "vignette of primitive America," where the hand of the white man had been largely removed from a self-regulating ecosystem. Yet in an important sense Yellowstone had become *more* primitive in recent weeks, more of a wilderness. The *wildness* of the wilderness had returned. Man again was engaged in a pitched and dangerous struggle to subdue nature.

And nature was in a very bad mood. Nature already had sent three dry windy cold fronts across the park—on August 6, 11, and 15—humbling the fire gods and shattering the last shreds of self-confidence left at Mammoth. Mysterious, inexplicable happenings were creeping into the equation. Contrary to the general expectations of fire behavior, fire perimeters were flaring up at night. Blazes were abruptly shifting course. Backcountry spike camps were reporting weird lights and eerie cackling from the burned mountains. Maybe it was just the cracking and falling of dead limbs and dead trees. Maybe the Great Spirit was angry. Maybe witches prowled the stunned woods. Or maybe it was only bears trampling through the bone-dry underbrush. Whatever it was, nobody got much sleep. Something was in the air.

Dan Sholly remembers having fallen into an exhausted slumber at midnight. He was up again three hours later, sitting on the edge of the bed and rubbing his one good eye. He pulled on his clothes and walked across the Mammoth compound to the operations center. Two of his top deputies, Rangers Steve Frye and Phil Perkins, already were there.

Every day was a battle for resources, and August 20 would prove no different in that respect. Mammoth's influence with Area Command seemed to ebb and flow in response to a mysterious bureaucratic tide, as Yellowstone Park fought to protect its turf, and Area Command fought to control the flow of resources in the Greater Yellowstone Area, and BIFC in Idaho fought to direct resources to a worsening fire situation all over the western United States. On August 19, the string had finally run out for BIFC: there were no more trained firefighting crews available. The national fire center activated an emergency agreement with the Department of Defense to quickly train and dispatch soldiers to the fires. Publicly, the move would be interpreted as Yellowstone's panicky response to the events of August 20, but the Defense Department call-up had been anticipated for weeks.

Sholly hurriedly conferred with Perkins and Frye in front of the huge fire map in the operations center. It would be two days before the Army could complete the crash firefighting course and fly in the troops. No help would arrive on August 20, and Sholly needed help, particularly on Clover-Mist. The fire, which he had been managing as incident commander since early August, was spotting down into the bottom of the Cache Creek drainage. "A well-developed fire in the creek bottom created the possibility that the southwest wind would blow the blaze right up the northeast line of Cache Creek, over Republic Pass, and into Silver Gate and Cooke City," Sholly later recounted. "Alternatively, Clover-Mist could blow straight north, jumping past Cache Creek and into the Soda Butte Creek drainage, and move on the gateway communities from there. Either way, it meant trouble."

Until August 20, the top priority for Area Command and Mammoth had always been somewhere else: first at Grant Village, then at Old Faithful, then briefly at the Fan fire, then back to the North Fork fire as it swept up the Madison Plateau and over Madison Junction. Now it was Clover-Mist's turn.

Sholly, Frye, and Perkins discussed the day's strategy as more people filtered into the operations center and quietly gathered around. Frye had been working for weeks without a break and Sholly ordered him to take the day off. "When dawn came," Sholly says, "we would reconnoiter Clover-Mist by helicopter and Perkins would work the fire as Operations Section Chief while I would go lobby Area Command for more resources. We had 300 firefighters and two helicopters stationed in the Lamar Valley above Cache Creek. I also had positioned two bulldozers at the Pebble Creek campground near the Soda Butte drainage, 5 miles northeast of the fire."

Sholly remained passionately opposed to dozer lines on Yellowstone property, but if Clover-Mist climbed north into Soda Butte drainage he was prepared to cut a wide swath across the creek and up toward the high ridge between Soda Butte Creek and Cache Creek, a ridge called the Thunderer. The two heavily forested drainages, and the ridge between them, appeared to be the battleground on which the fate of Silver Gate and Cooke City would be decided.

Clover-Mist
5 A.M.–10 A.M.

As dawn came on August 20, seven major fires were burning in the Yellowstone region—Fan, Hellroaring, Storm Creek, Clover-Mist, North Fork, Mink, and Snake River. Out of the seven, two would emerge as most important. One was the North Fork fire. The other was not a single fire but rather a complex of fires—Clover-Mist, Storm Creek, and Hellroaring—that would burn in and around the mountains, meadows, creeks, and roads surrounding Yellowstone's great northern range, the Lamar Valley.

Somehow, the Lamar Valley had managed to remain a back page in the catalogue of Yellowstone's wonders. Less-traveled than other parts of the park, it is a pastoral medley of high country glades, abundant wildlife, and distant snow-capped peaks. The trapper Osborne Russell named the area "Secluded Valley," noting in his journal of 1836 that he had traded there with "independent and happy Natives," and adding that there was "something in the wild romantic scenery of the valley which I cannot describe." Forty-nine years after Russell's visit, Secluded Valley was renamed in honor of Lucius Quintus Cincinnatus Lamar, Secretary of the Interior under Grover Cleveland—politics once again triumphing over poetry.

Bliss it had once been for Dan Sholly to come zooming in by helicopter over that wild romantic land! In less than five weeks his life had been turned upside-down. Sholly remembers grimly staring out at the smoky dawn landscape on August 20 as Curt Wainwright steered the Llama into a landing zone in a dry meadow by the confluence of the Lamar River and Soda Butte Creek. Now the whole thing was like some horrible dream he couldn't escape.

Sholly's crews, all Yellowstone volunteer personnel, were standing by in the meadow. For two weeks Clover-Mist had been creeping down into Cache Creek. "With a little luck, I felt we could get some crews in and build a line across Cache Creek that would hold the fire off," Sholly recalled. "We had ordered two big Chinook helicopters to ferry the firefighters in, but the aircraft hadn't arrived."

While they waited for the Chinooks, Phil Perkins dispatched his division supervisor—an Operations Section Chief's lieutenant, responsible for a single area within a multi-area fire—and two helitack firefighters to Cache Creek. Wainwright dropped the three men off and headed back to the Lamar Ranger Station for refueling. The division supervisor's mission was to survey the landing area and prepare to receive and deploy the troops the Chinooks would bring in.

From his position by Soda Butte Creek, Perkins recalled, he could not see south over the hill to Cache Creek. He paced uneasily around the landing zone, waiting for the Chinooks. "This was a critical operation," Perkins says. "Our structures at the Northeast Entrance to the park and the gateway communities now unquestionably were within reach of Clover-Mist." Perkins glanced to the northwest, where a patch of clear blue sky earlier had met the morning.

" 'Holy smokes!' I thought. 'What was that!' " An enormous convection column, "twisted up like a gigantic braid of rope," had suddenly appeared, climbing to 30,000 feet high. It was the Hellroaring fire, nearly 15 miles away. Perkins looked at his watch: just past 9 A.M. The experienced firefighter had never seen anything like it so early in the morning.

Then he felt the hot breeze on his face and whirled

around to face Cache Creek. A convection column was starting to build there too.

Perkins' radio squawked to life.

"The fire is picking up here, Phil," his division supervisor said, as Perkins recalled it.

"Can you contain it?" Perkins asked, knowing and dreading the reply.

"Negative, negative," came the laconic response. "Can you pull us out?"

Perkins contacted Wainwright. He was almost finished refueling at the Lamar station and would lift off in two minutes. But that wasn't fast enough. "A second Llama, piloted by Doc Herzberg, was waiting inline for fuel. He took off for the three firefighters," Perkins recalls "I ordered him to get them out ASAP."

"Uh, Phil." The division supervisor radioed in. "The fire is about to run over us. We'll have to deploy shelters in a couple of minutes."

"Not again!" Perkins remembers with a rueful chuckle. "Every shelter deployment triggered a mandatory investigation." Sholly already was getting nitpicked to death over the incident at Calfee Creek.

"Helicopter is on the way," Perkins responded. "Deploy if you have to."

The wind was picking up. "I knew that air operations would be grounded soon," Perkins said. Just then Herzberg blasted overhead, dropped into Cache Creek, and pulled out the three firefighters.

The plan to tie the fire off at the western end of Cache Creek was abandoned. Sholly flew back in for a quick conference. All hell was breaking loose throughout the park and they could expect no further resources from Area Command today, Sholly reported. The wind began to whip around them and they had to raise their voice to be heard.

"At all costs," Sholly yelled to Perkins and the grim-faced men gathered in the smoky meadow, the convection column behind them climbing out of Cache Creek, "we have to keep the fire out of Soda Butte Creek!" The drainage was a virtual wind tunnel pointed right at the Northeast Entrance, Silver Gate, and Cooke City.

Perkins remembers nodding, silently listening to Sholly's instructions. The wind whipped at the high dry grass around them. "Sholly authorized the use of bulldozers at that time," Perkins says. "But the fire had a mind of its own."

Silver Gate
10 A.M.–1 P.M.

It didn't much matter to Hayes Kirby whether the Clover-Mist fire came the long way up Soda Butte Creek and through the Northeast Entrance or the short way up Cache Creek and over Amphitheater Mountain. "It was coming to burn out Silver Gate, that's all I cared about," Kirby remembers. "It was god-damn bureaucratic arson!"

Kirby, the Texan who owned property in Silver Gate and Cooke City, was a former Air Force pilot with a mysterious past and weighty Pentagon connections. He had been trying to get action on the Clover-Mist fire for weeks. That day, Kirby paced around the bleak-eyed totem pole outside the Grizzly Lodge, his motel beside the forest at the edge of Silver Gate. "There was a poison in the air," Kirby said, "a sickness, and it wasn't just the goddamn smoke. I remember around noon glancing over my shoulder at the convection column rising behind Amphitheater Mountain. The son-of-a-bitch was heading right for us!

"That sickness in the air, I sure as hell recognized it," Kirby recalled. "It wasn't just the goddamn fires. I'd seen it before in Vietnam: no one knew what the fuck they were doing. The thing had gotten away from them. The beast was out of control."

For Kirby, Silver Gate was the last best place, a place where the cowboy spirit was still alive. He was the son of Texas oil millionaires. "People hereabouts are like my people had been in Texas before they were changed by staggering amounts of money," Kirby said. "Primitive people, in the best sense of the word. Mountain people. Ornery sons-of-bitches."

Kirby had tried to warn those *other people*. "I *begged* Bob Barbee and Dan Sholly to fight Clover-Mist," Kirby recalls, "but no, they said, the fire was 'small,' 'insignificant,' 'in prescription.' " Kirby had flown his own plane over the fires and contradicted park reports. He had stood up at public meetings and "made a goddamn fool of myself," shouting at his neighbors, *"Don't you people understand that they're gonna burn you down! We need action! We need those fires put out NOW!"*

But now it was too late. By 1 P.M. on August 20 the smoke column behind Amphitheater Mountain was boiling higher and ash was floating down from the slate sky and the wind was shaking the trees. "Holy shit!" said Kirby. "This was it! Here came that small insignificant fire the park had been playing down all summer! Yellowstone National Park had won again. It's like living next door to Cuba! They had the absolute authority to burn Yellowstone Park to the ground and by God it looked like they were gonna do it!"

Kirby didn't hate Bob Barbee and didn't believe Barbee was "part of some deep dark plot to destroy the gateway communities, as some people did." As for Sholly, Kirby seemed to respect him, as one former military man to another. "The son-of-a-bitch was a likable man," Kirby says, "a man who exercised power in an impressive manner. But Barbee and Sholly were creatures of a political sickness. Yellowstone National Park had been turned into a laboratory for the idea of a self-regulating ecosystem, and that explains the running defense of the park by special-interest groups such as the Greater Yellowstone Coalition."

Kirby remembers angrily watching ash drifting down on Silver Gate's one paved street. "Environmentalists talk of 'returning' Yellowstone to its 'original, pristine condition,' " Kirby says. "But what is pristine? Following the environmentalist's logic, 'pristine' meant no people, no economic development, removal of the gateway communities, removal of the nonindigenous buffalo, the exotic grasses in the Lamar Valley, certain types of trout—the list goes on and on."

Kirby decided he wouldn't leave town. He would wait for the fire to come to him and continue to fight what he called the "political sickness."

"I knew the political sickness all too well," Kirby says "Barbee, Mott, Hodel, the forest supervisors, Area Command, the media, the politicians, the special-interest groups—all of them were serving their own individual ends, while the god-damn thing keeps growing and growing until finally it takes on a personality of its own. I'd seen the exact same thing in Vietnam and Angola and Nicaragua. Everybody was just orchestrating their own little part of the affair: *Nobody was really running the goddamn show.* Meanwhile, the bastard just keeps getting bigger, until one day it explodes."

North Fork
1 P.M.–2 P.M.

It was a day for watching the sky. Phil Perkins watched it from the Lamar Valley. Hayes Kirby watched it from Silver Gate. Dave Poncin watched it from the fire camp at Madison Junction, watched the dark columns rising from the woods and spreading great boiling gray heads north by northeast, watched the little dust devils and whirlwinds skirting across the parking lot, the forest swaying. By 1 P.M. the winds were climbing up over 30 miles per hour. The forest began to burn hard.

Poncin had over 1,000 firefighters positioned across the Madison Plateau, Madison Junction, Canyon Village, West Yellowstone, and the Norris Geyser Basin. But Poncin was a practical and careful man, and he knew there was little the firefighters would be able to do today. The incident comman-der would try to hold what lines he could and keep his people out of harm's way.

By 1:30, they were coming into the peak burning period. Poncin had just returned from a town meeting in West Yellow-stone where Bob Barbee had been raked over the coals. As in

124   Silver Gate, the locals in West Yellowstone were in an ugly mood. Rumors were flying that Dan Sholly had put a plain-clothes security detail around Barbee in response to assassination threats.

Poncin had noticed that the park officials were not enthusiastically pushing the positive ecological aspects of fire anymore. That was smart. The last thing the people in West Yellowstone wanted to hear about was the stimulating role of fire in the ecosystem. Barbee had assured them that everything possible was being done to fight the North Fork fire, now 9 miles east of the town.

In a hurried conference in the parking lot after the town meeting, Poncin explained his strategy to Barbee. "I wanted to flank the main fire front and take action to protect West Yellowstone and the structures at Norris Geyser Basin, 14 miles up the Gibbon Canyon road from Madison Junction." Poncin later explained. "I wanted to drive a bulldozer line south from West Yellowstone along the park's western boundary to shield the community, and deploy structural protection teams and engines around the buildings and bridges at Norris." He needed Barbee's consent for the dozer use. The superintendent immediately agreed.

By 2 P.M. Poncin was back at his base camp and studying the latest intelligence. The fire was raging in Gibbon Canyon, the same place where Lt. A. F. Boutelle's troops had labored in 1890 without shovels and buckets. Little had changed in a hundred years. Resources still were short.

The fire funneled into Gibbon Canyon, turning the woods into a wasteland of smoldering stumps and ash. All Poncin could do was marshall his resources, work on the fire lines, and try to protect structures and power lines. He wanted to contain the blaze's southward movement and drive his fire line east to link up with the road between Norris Geyser Basin and Canyon Village. One day soon, he knew, the wind would die, the humidity would rise, or a good rain would come, and they would be there, ready to strike back at the fire that had been kicking them halfway across Yellowstone National Park.

Mink
2 P.M.–5 P.M.

Suddenly, there were more new ignitions. At 2 P.M. the big wind tumbled an aspen across a power line east of Jackson Hole, Wyoming, starting the Hunter fire in a boundary area of Grand Teton National Park and Bridger-Teton National Forest. Fanned by the storm, the blaze grew an acre a minute, threatening nearby homes and ranches. Grand Teton National Park scrambled a crew and attacked Hunter within twenty minutes, but it was useless. The fire rapidly moved out of the park and into heavy fuels at Ditch Creek, in the Bridger-Teton National Forest. By 3 P.M. the call had gone out for another Type I team.

The scenario would repeat itself less than a hour later at the opposite end of Grand Teton National Park. Twenty-seven miles north of the new Hunter fire and only 4 miles south of Yellowstone Park's southern border, a blistering curveball of wind swept in from the southwest and blew another aspen across another power line, igniting the Huck fire. The blaze immediately leaped over the John D. Rockefeller, Jr., Parkway and flashed northeast toward the Flagg Ranch resort, triggering a hasty evacuation. Driven by 30 mile-per-hour winds gusting up to 60 miles per hour, Huck slammed into dense lodgepole pine and heavy dead fuels, exploding across 4,000 acres in two hours. By 5 P.M. a call had gone out for yet another Type I team.

Twenty miles to the east, Incident Commander Dale Jarrell agreed to take over the Huck fire. Jarrell was back for a second time as IC of the Mink blaze. The incident commander's first move was to send scouts and a strike team to help ranch hands secure the structures at the Flagg Ranch. He also contacted the U.S. Forest Service's Region Four dispatch center in Ogden, Utah—thereby bypassing Area Command and BIFC—and ordered another 200 firefighters, six helicopters, and four engines. Apparently, there still were a few resources around if one knew where to look.

A lot had happened since Jarrell had transitioned off the Mink fire on July 31 and turned it over to a Type II team. The fire had been declared "100 percent contained/confined" but soon revealed itself to be creeping out of the confinement boundaries on its north end, inside Yellowstone Park. In those optimistic early days of August, Barbee and the rest of the Yellowstone high command resisted pressure for an aggressive attack of the fire. Mink slowly spread north, into Cliff Creek, Mountain Creek, and east from Mountain Creek toward Howell Fork.

On August 9, the fire was turned over to a second Type II team. Light rains were falling on the southern parts of Yellowstone and the situation appeared stable. But by August 14, the weather had turned hot again. Mink moved another 2 miles north to Trapper Creek and spot fires increased on the potentially troublesome south end of the blaze, near the big timber blowdown and the homes of Buffalo Fork. Two days later, lightning started a new fire one mile west of the Mink fire, near Emerald Lake in the Teton Wilderness. By August 18, the Emerald fire was gaining strength and threatening to move into Yellowstone Park.

The Mink fire, on the other hand, was cutting a northeastward path across the corner of Yellowstone and was poised to move *out* of the park at Mountain Creek and return to national forest land at Howell Fork. Bridger-Teton National Forest Supervisor Brian Stout wanted Mink stopped. The call went out for a Type I team and by August 20 Dale Jarrell was back on the job. By 5 P.M. he had assumed responsibility for the Emerald, Mink, and Huck fires.

But stopping anything on August 20 was out of the question. The hot hard wind was mounting and the sky was filled with a rain of fire. The last twelve hours, from dawn to dusk, had brought major advances on all fronts in the Yellowstone area, from as far south as the Hunter fire near Jackson Hole to as far north as the Hellroaring fire in the Gallatin National Forest. In the heart of Yellowstone Park, at Madison Junction, Dave Poncin glumly listened to his scouts' accounts of North Fork raging along the Gibbon Canyon road.

Bob Barbee hadn't even been able to make it back to Mammoth after the West Yellowstone meeting and that afternoon he could be found helping out with the evacuation of West Thumb at Yellowstone Lake. Phil Perkins was huddled with Dan Sholly, planning to march the park firefighters up to the 10,000-foot ridge line of the Thunderer for a final attempt to keep the Clover-Mist fire from blowing out of Cache Creek, over the divide, and into Soda Butte Creek. At the other end of the Soda Butte drainage, thick smoke and ash covered the narrow two-lane highway out of Silver Gate through the forest to Cooke City, Cooke Pass, and the Clarks Fork Valley. Down the highway in Red Lodge, 70 miles from the park, the sky all day had been a burnt orange, tricking the streetlight on at noon. "It looked," one resident recalled, "as if the Apocalypse had come."

And there was yet another nasty surprise lurking in the night. Maybe it wasn't Clover-Mist that would engulf Silver Gate and Cooke City after all. Thirteen miles north of Cooke City, a coyote fire, trickster, terrorist, was creeping through the woods. It had been the first fire of the summer, a forgotten blaze slumbering deep in the wilderness of the Custer National Forest. Now it was about to come hurtling out of nowhere, bewildering the fire gods with its perverse behavior and bringing destruction right to the doorsteps of the two mountain hamlets. Storm Creek was on the move.

Storm Creek
5 P.M.–8 P.M.

On August 19, Beartooth District Wilderness Rangers Dan Hogan and Tom Alt had left headquarters in Red Lodge, picked up a Forest Service pack horse, met Ranger Steve Studer and hiked into the mouth of the Stillwater drainage, heading up into the steep canyon country to install fire-protection sprinklers on several wood bridges used by backpackers and outfitters. For most of the previous nine weeks, Storm Creek had been a low-priority fire. The Type I team called in by Beartooth District Fire Management Officer George Wel-

128   don on July 4 had built fire lines, conducted burnouts, and effectively prevented any northward movement of the fire out of the forest and onto the campgrounds, homes, and mining site beyond the wilderness boundary. Fire-behavior analysts reported that the prevailing southwest winds, as well as natural fire breaks—cliffs, rock slides, avalanche chutes, and steep areas of sparse fuel—would block Storm Creek's southern movement toward Yellowstone Park and the gateway communities. In other words, Storm Creek was contained. The Type I team departed on July 10, turning the fire over to the Custer National Forest.

On August 18, after five weeks of minor activity, the fire started backing south into heavy fuels near Wounded Man Creek. A helitack crew had cached most of a sprinkler system under one of the bridges at the creek. Hogan and Alt's mission was to bring in the rest of the hose—courtesy of Star, a twenty-five-year-old chestnut gelding—rig the gravity-flow sprinklers over two bridges in the Wounded Man drainage, and evaluate the structural protection needs of a ranger cabin near Big Park meadow, a 30-acre pasture along the Stillwater River near Storm Mountain.

It was Dan Hogan's first year with the Forest Service. That spring he had joined as a seasonal employee, completing the one-week training course for all new employees and receiving his basic "red card" certification, a wallet-sized document qualifying him for firefighting work. Although not an experienced firefighter, the Montana-born wilderness ranger was a savvy woodsman and he didn't like what he saw as he hiked into the Stillwater on August 19. Spot fires flickered all around him in the forest and a large column of smoke was climbing out of Storm Creek. "I felt uneasy walking by the spot fires," Hogan remembers. "It didn't seem natural to let them burn without doing anything."

Around 9 A.M. on August 20, while working on one of the bridges, "we noticed a big column of smoke boiling up from the direction of Hellroaring Creek, about 20 miles to the west," Hogan says. It was the same smoke column that Phil Perkins, roughly 20 miles south of them at Soda Butte Creek, was watching. Tom Alt, like Phil Perkins, took the early-morning emergence of the column as a distinctly unfriendly sign. They hurriedly rigged the hose and sprinklers. Alt then departed to patrol the area while Hogan waited by the bridge.

In the afternoon, Fire Management Officer George Weldon and Ranger Blake Chartier hiked in and met Hogan and Alt by the bridge. The fire, they reported, was active in the valley below, but was still a good 4 miles away. Weldon burned out the high grass around the bridge and the others worked to clean the sprinkling system: the ash-filled air was fouling the water and clogging it.

Around 7 P.M. the four men—the fifth, Studer, had moved upcountry to check out a report of some campers in a closed area—walked back to the cabin at Big Park meadow. Alt and Weldon, the two senior officers, discussed strategy. An overflight reported that although the fire was still burning hot it was at least 3 miles downstream from the Big Park Ranger Cabin—not an immediate threat, in other words. The next morning, Alt and Weldon decided, they would drop smoke jumpers and pumps into the meadow to protect the cabin.

"It was getting dark," Hogan recalled, "when Alt walked away through the woods to turn Star out in the meadow for the night. Chartier and I went into the cabin and put water on the stove to start dinner. In the distance was a rumble, like the sound of a freight train. I remember that Weldon was pacing around outside the cabin, watching the glow in the sky, listening, bending down now and then to pick up a bit of debris that had floated through the air." Suddenly it grew calm. "Chartier and I sensed something was happening and went to the door of the cabin. Weldon looked over at us and said: 'We gotta get the hell out of here.' "

Just then Tom Alt came out of the woods with Star. "Alt had come to the same conclusion as Weldon," Hogan remembers. " 'We've got to get going,' Alt said to us. 'Head to the meadow.' "

Hogan grabbed a water bottle, a sweater, and a lead rope for Star. They threw a saddle on the horse so the embers wouldn't burn his back. Weldon snatched up a portable pump and some flares. "The rumble had turned to a roar as we ran through the woods to the edge of the meadow," Hogan says.

Alt and Weldon hurriedly discussed their plan. Time was running out. "Meanwhile," Hogan says, "I was trying to read the instructions on my fire shelter. Suddenly I was feeling a pressing need to understand how the thing worked. The wind was just *screaming* up the valley. Alt and Weldon lit the flares, bent to the dry grass, and—BOOM!—the meadow exploded, as if they had thrown a match into a puddle of gasoline." Within sixty seconds the flames had raced across the 30-acre pasture and into the forest. Behind them, the fire was hitting the woods.

They moved into the black. A driving rain of ash and embers enveloped them. The furious wind, gusting up to 70 miles per hour, sent Hogan staggering. He grabbed Star's lead rope to steady himself. The horse jammed his muzzle against the ranger's chest and kept it there. They huddled together in the middle of the burned field. They could see little now— bandannas covered their faces, protecting them from the fiery hail—and probably they wouldn't have wanted to see it any- way: the wall of flame advancing toward them, the mountains consumed in fire, death dancing on the wind.

Clover-Mist
8 P.M.–11 P.M.

South of Storm Creek, at that moment, Phil Perkins was hik- ing two crews of Yellowstone firefighters up the Thunderer. All day and into the early evening the Yellowstone teams had dug line in the Lamar Valley. The winds had whipped at them and they could hear the fire's awesome rampage just over the hill, consuming almost 10 miles of Cache Creek between 9 A.M. and 8 P.M.

During a lull in the windstorm, Wainwright had flown in and taken Perkins and Sholly up for a look at the blaze. Clover-Mist had burned up Cache Creek beyond the 9,000- foot elevation and was at Republic Pass, only 3 miles from

the dense lodgepole forest surrounding Cooke City and Silver
Gate. Hastily evaluating the situation, Perkins and Sholly saw
at least three distinct paths the fires could take to the gateway communities. It seemed to them that the most likely scenario was that Clover-Mist would spot north across Thunderer ridge, then be funneled up Soda Butte Creek, blowing
over the structures at the Northeast Entrance and along the
road to Silver Gate and on to Cooke City. Or the fire might
spot northeast across the grassy fuels and rock fields at the
top of Republic Pass and move down Republic Creek onto
Cooke City. Or the fire might burn on a more easterly course,
threading out the drainages of Pilot Creek and Onemile
Creek.

   "In one sense," Perkins recalls, "we were fortunate. Thunderer ridge, running from southwest to northeast, had contained Clover-Mist in Cache Creek. If the Thunderer had run
on a straight east-west course, for example, the strong winds
blowing to the northeast would have thrown the fire right over
the ridge. Instead, the ridge channeled the fire. But the danger
that Clover-Mist would come over the ridge into Soda Butte
Creek remained. The dozers were standing by at Pebble Creek
campground. Sholly, however, still did not want to use them.
He thought that if we could put two crews up on top of the
Thunderer that night, we could hold Clover-Mist on the Cache
Creek side of the ridge."

   It had been a very long day and it was going to be a very
long night. There was no time for dinner—the Yellowstone
crews gathered some rations, some water, organized the gear,
went over the plan, and marched. Escape routes and safety
zones were identified. Scouts were posted around Cooke City,
Silver Gate, and Republic Pass. A break in the weather
allowed a retardant bomber in.

   By 10 P.M., they were up on top of the Thunderer—
named, someone suddenly remembered, for the uncanny way
it attracted thunderstorms. "The haze parted for a moment,"
Perkins says, "and the view was enormous." They were surrounded by a night world of flame-clad hills, burning hori-

zons, and isolated trees flaring in patches of darkness. To the north was Storm Creek; to the west, Hellroaring; to the south and east and right at their feet, Clover-Mist.

The foul air closed in around them again and they fanned out across the ridge line, chopping at the thin soil with hand tools. It was unfamiliar territory, night territory, and Perkins posted extra lookouts. Headlamps fixed to helmets stabbed through the smoke, searching out the glowing cinders that drifted up from the drainage below. "Midnight came," Perkins says, "and we were still chasing spots and digging line." Spotting embers would find their mark and trees suddenly would torch up, abrupt apparitions in the hellish darkness.

It made you nervous, as some of the men up on the Thunderer would later recount. It made you want to run away, but there was no place to escape from it. The work made you tired, so tired that you wanted to curl up and take a little nap, but that wasn't at all smart—that was the way you woke up dead. The long hours and the endless weeks of fire made you want to give up, to retreat into the black, to walk away through the smoking ruined woods and never look back. But then you would be leaving everyone else behind, and you couldn't do that either.

Mammoth Hot Springs
Midnight

They met at midnight at the Mammoth operations center. There was no panic—there might well have been, as initial reports were estimating that over 100,000 acres had been eaten by the flames in a single day. Final estimates would put the day's fire growth at 160,000 acres, running the total fire perimeters in the Yellowstone area up to 480,000 acres.

Participants at the midnight meeting recall an air of dazed numbness and a dogged determination to hang on to whatever was left. The top Yellowstone brass all were there. They had to make some big decisions fast. Through it all, through almost two months of fire, Barbee had managed to keep the park open, selectively closing roads and campsites as a fire rolled into a threatening position and reopening them as soon as the blaze moved off. Visitation had held steady in June and July as tourists, reassured by Yellowstone's message that the fires were both unthreatening and part of nature's process, decided not to cancel long-planned vacations. Visitation had plummeted in August, however, and the summer was turning into an economic disaster for TW Services, the park's chief concessionaire. The midnight meeting would have to decide: Should they, finally, close the park? Should they, in effect, surrender?

"I remember just staring at that big fire map on the wall," Barbee says. August 20 had been a disaster, one for the record books. Already firefighters had dubbed it "Black Saturday" and were comparing it to the "Big Blowup" of August 20, 1910, exactly seventy-eight years earlier. "It had been a black Saturday down in the southern half of the park, that was for sure," Barbee recalled. "I'd seen it myself. It looked like we would have to close Grant Village and the Lewis Lake campground again. It also appeared that all the fires in the Snake River complex were about to burn together." The combined Red-Shoshone fire in the last forty-eight hours had moved 4 miles to the northeast, burning up Beaver Creek, across the Continental Divide, and into the Continental fire. Further south, the Falls fire had been running northeast at a mile a day for three days. At that rate, it would burn into the south end of Red-Shoshone-Continental in seventy-two hours. Nearly 600 firefighters under a Type II overhead team were attempting to control the Snake River Complex, which now covered 70,000 acres.

Four miles below the South Entrance, the Huck fire had missed the Flagg Ranch and was on a course that would bring it into the park, and possibly together with both the Snake River and Mink fires. Total personnel on the Huck-Mink complex was nearing 300 and Incident Commander Dale Jarrell had asked Yellowstone for permission to put a camp above the

134   northern head of Mink, deep in the backcountry in a fragile meadow surrounded by grizzly bear habitat.

"I was warned that the bears would zoom right in on the firefighters, and that garbage and scores of people and equipment would ruin the meadow," Barbee says. He approved the request, but continued to insist on light-hand practices and daily removal of garbage to forestall bear encounters.

Further south, near Jackson Hole, the Hunter fire had grown to 2,000 acres. Jarrell had sent four crews down to help battle the blaze and a new Escaped Fire Situation Analysis was being prepared, calling for more resources and total control of Hunter within forty-eight hours.

"We would have to prepare a number of our own new EFSAs that night," Barbee recalls. " 'Escaped fire' was the name of the game."

At the North Fork fire, there had been some spotting on the southwest flank; most of Dave Poncin's lines had held. But there was no controlling the fire's free-burn to the north and east. On August 18, North Fork had been 69,000 acres; now it was over 91,000 acres.

"With the head of North Fork past Madison Junction," Barbee says, "I thought maybe we could keep the road from West Yellowstone to Old Faithful open." Some of Sholly's rangers, responsible for enforcing the roadblocks and dealing with traffic flow, argued that now was the time to keep all visitors out of the park. The fires were just too dangerous, too unpredictable. They needed to keep the roads clear for the speedy movement of resources.

Some of the park scientists at the meeting agreed. Even the strongest supporters of natural fire were starting to give ground. Fire behavior had become almost inexplicable. For example, reports filtering in to Mammoth were indicating that the Fan fire, in the northwest corner of the park, had started to move *south*, against the prevailing winds. Even with 700 firefighters on the blaze, there was no way to quickly and effectively respond to such odd behavior. But that wasn't even half the story. Twenty miles east of Fan, the Hellroaring fire had just run *5 miles north*—north!—consuming the headwaters of Hellroaring and Middle Fork creeks in an eight-hour, 14,000-acre run, between 9 A.M. and 5 P.M. Twenty miles further east, between 6 P.M. and 10 P.M., Storm Creek had run 7 miles in four hours—7 miles *south*, in the *opposite direction* of the Hellroaring run.

No one could say for sure what was going on in the northeast sector of the park. Was the gigantic Clover-Mist fire, now at over 150,000 acres, with its northernmost point at Republic Pass only 14 miles from Storm Creek's southernmost point outside of Horseshoe Canyon, creating its own immense independent weather pattern? Was all that superheated air hurtling upward acting like a giant vacuum, sucking Storm Creek into Clover-Mist? Was that the explanation for Storm Creek's southern lunge?

If that was the case, then Silver Gate and Cooke City were in the center of a tightening ring of fire. Not only had Clover-Mist's northern flank burned up Cache Creek to within 3 miles of Cooke but its eastern flank had blown nearly 3 miles out of Bootjack Gap, just as Incident Commander Curt Bates had warned almost three weeks earlier. Clover-Mist thus was in a position to threaten the gateway communities from the south, while Storm Creek was coming down from the north. Phil Perkins, for his part, had held the fire at the Thunderer.

Bob Barbee turned to the others. In the distance he heard a helicopter approaching. At last the winds were abating. They would leave the entrance from West Yellowstone open to visitors and evaluate the other closures on a daily basis. Technically, Yellowstone Park would remain "open." As for Clover-Mist, they would immediately prepare a new EFSA to raise it to a high-priority fire and press Area Command for a Type I Incident Management team and increased resources. The new EFSA would recommend full control, with extensive handlines, burnouts, and air operations.

As Barbee spoke, similar decisions were being made at Area Command, in the Shoshone National Forest for the Clover-Mist fire, in the Gallatin National Forest for the Hellroaring fire, and in the Custer National Forest for the Storm

136   Creek fire. Nature had taken its course. In the hurly-burly of
Black Saturday, the battle against the fires had been lost, but
natural regulation had won. Fire was functioning in the
ecosystem.

Storm Creek
After Midnight

The ground retains heat long after a fire has passed. Sometimes invisible fire will smolder in the duff, flickering to life
hours or even days after the fire front has vanished. That's
why a fire line is cut down to mineral soil. That's why a firefighter will take his glove off and work his hand deep into the
black, searching for the warmth of hidden fire.

Dan Hogan, Tom Alt, George Weldon, and Blake Chartier
stood in the meadow until 1 A.M. The packhorse, Star, stood
there too, unpanicked as the wind screamed and shot tracers
of fire. At a Christmas party later that year, the Beartooth District would award Star his own red card, qualifying him as a
genuine firefighter.

It had been safe in the big meadow once Weldon and Alt
torched it off. It was frightening, but once the grass was gone
only panic would have killed them. Long after the front
passed, Tom Alt wandered over to the river and found a spot
of green, a patch of grass the blaze had hopped over. He called
to the others. "It was as good a place as any to wait for dawn,"
Dan Hogan remembers.

They huddled on the grass and tried to sleep. A million
eyes of fire gleamed at them from the surrounding mountains
and the night was filled with the sound of cracking limbs,
popping stumps, and falling trees. The Ranger Cabin was
gone, reduced to a slab of smoking cement. Hogan peered at
the others. "They were completely black," he says, "covered in
ash from head to toe. I hadn't expected quite that much excitement from a Forest Service job. I love the mountains and had
figured the job was simply a way I could stay in them. Instead
it was like: Join the Forest Service, see the wilderness, get
burned up. I won't soon forget that particular Saturday."

The night began to pass from fire to ice. "It gets cold in the high country, even in August," Hogan says. "I noticed a log smoldering a few yards away, its cavity all scooped out and glowing with hot coals." Hogan slowly rolled over onto his hands and knees—he was tired, dead tired—and crawled out of the green grass and into the black, his hands raising little clouds of ash as he moved to a warm spot beside the log. Star snorted and pawed at the ground. The others looked up. One by one, Alt and Weldon and Chartier followed Hogan over to the log and snuggled down into the warm ashes. That was better. Warmer. Dawn would come sooner now.

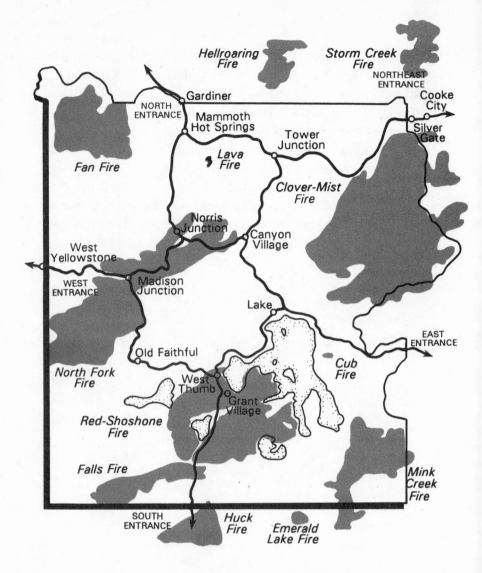

YELLOWSTONE FIRES
August 21

After Black Saturday. An additional 160,000 acres have
burned. Total fire perimeters now cover 480,000 acres.
Storm Creek and Hellroaring fires move south.
North Fork fire divides into multiple heads.

BOOK FOUR

Paradise Lost

Accuse not Nature, she hath done her part;
Do thou but thine!

—MILTON, *PARADISE LOST*

XIII

North Fork

PARK POLICY DEVASTATING

. . . Reports have come as from a battlefield. Yellowstone National Park has come under fire, and tourists and park workers have been evacuated as soldiers were evacuated from Saigon, fleeing a tide they cannot turn.

And in Billings, the sky is gray day after day, gray with soot, gray as the chances of stopping the fires raging across the park.

[What we need now] is an explanation from Park Superintendent Robert Barbee about how all this could have happened.

We know the standard response: the park has a fire policy to ensure the area grows and develops as it would have in a pristine state. If man interferes with the process, the park is no longer a park.

Under normal circumstances that policy makes good sense. But conditions at the park this year are anything but normal, and the policy comes up as dry as Montana. Barbee and company should have seen that. The fact that they didn't is a matter of grave concern.

We are in the midst of a record drought. Pine needles

are brittle as glass and drip like gasoline from trees.

One match, one flash of lightning, a park policy that says burn, baby, burn, and the park was aflame.

In response, the Park Service did nothing but spray policy at the blaze until it was too late. Then, when the war was lost, an army of firefighters was called in.

This fiasco is riddled with questions, and it's not too late for Congress to demand to know why Barbee rode a dead policy into hell.

Billings Gazette editorial,
August 28, 1988

One of the principal effects of Black Saturday was to bring the return of the national media. Although the regional press and television—led by detailed and often sharply critical coverage from the *Billings Gazette* and the *Denver Post*—had closely followed the story since early July, the national press corps by the first few days of August had drifted away in the aftermath of the Shoshone fire's run at Grant Village, the Hodel press conference at Old Faithful, Area Command's optimistic assessments, and the park's repeated assurances that the fires were nature's way of doing business.

NBC, for example, had filed its last fire story on August 5, a two-minute segment on fire's natural role in the ecosystem, featuring Superintendent Barbee and Chief of Research John Varley. A week later, NBC's *Today* show aired a long piece by correspondent Jim Fowler on reintroducing wolves into the Yellowstone ecosystem. Wolf reintroduction had been the major issue on the table for environmental groups before the fires began. Little mention was made of the fires. Similarly, the story all but vanished from the national print media. The pacesetting *New York Times* had last filed a ten-paragraph wire story on regional forest fires on August 2.

Black Saturday changed all that. Within forty-eight hours, according to records in Yellowstone's Public Documents Center, the networks, the leading dailies, the newsweeklies, CNN, and dozens of other newspapers, radio and television stations had posted reporters and crews to the area. Joan Anzelmo, the Yellowstone public affairs chief, doubling as media coordinator for Area Command, found that the press storm she had survived in late July was merely prelude to the howling hurricane of reporters clamoring for interviews and fire footage in late August. The press, moreover, were now in a surly mood. Many reporters felt they had been bamboozled by Yellowstone in July and were irritated at missing August 20, one of the most dramatic days in the history of wildland fire-fighting. From August 22 to the end of the fires, Yellowstone would be besieged by a hostile press.

"It was a nightmare," Joan Anzelmo recalls. "The fires were spread out so far and moving so fast that I was having trouble getting timely information. The phone or radio link to a fire camp would suddenly go down. Or the computer to BIFC would black out. Or one of my information officers working liaison at an incident command post would suddenly be missing in action—out searching for a photo op with some pushy jerks from newspaper X and network news affiliates Y and Z. Meanwhile, literally almost every half hour NBC and CBS and ABC would be calling from New York, demanding to know: 'Where are the fires now? Why are you letting them burn? Why aren't you putting them out? We need to know *now!*' Well, we *were* fighting the fires! Didn't they know that we had announced a moratorium on the natural fire policy? Didn't they read the press releases we were putting out every day? And I could have *killed* some of those reporters! Day in and day out, they were airing clips of flaming forests and irate locals like Hayes Kirby blaming the Park Service for 'burning down Yellowstone.'"

Anzelmo, Sholly, and others in the park were growing increasingly bitter about the press. "The coverage during the summer was pure sensationalism," Anzelmo says. "Bob Barbee was being turned into a villain, and the media were fueling the public perception that Yellowstone Park was in danger of being 'destroyed.'"

In Washington, cabinet officers, the regional congressional delegations, and high-ranking members of the Park

144   Service and Forest Service all began vigorous efforts to spin the coverage their way and exercise damage control, making hasty plans to visit the area. Area Command would soon find itself spending hours of precious time dealing with VIP visits.

The first beneficiary of this attention was West Yellowstone. The press flocked to the gateway community. Thanks to the media, everyone in the United States could now see the North Fork fire moving through the forests about 6 miles away. In a magnanimous gesture, a group of Mormon ranchers from Idaho banded together and went to west Yellowstone, where they proceeded to install a sprinkler system around the town. Politicians demanded action. Area Command was forced to make a visible response to media and political pressure, according to Forest Service officials. It diverted resources to deal with the real—if not at that moment critical—threat to the town.

For Dave Poncin, the critical threat posed by North Fork was not to West Yellowstone in the west, but to Canyon Village in the east. On Black Saturday, the fire had ended a 5-mile superheated up-canyon run at the cliffs of Secret Valley, a picturesque vale north of Gibbon Falls. The Gibbon Picnic Area, a favorite of park visitors, had been reduced to a charred landscape of black poles.

On Sunday, August 21, Poncin's pursuit of the North Fork fire resumed. "I was beginning to feel like Ahab," Poncin recalled, "chasing the great white whale." His epic handline along the northeast flank of the fire was closing in on the head of the blaze above Gibbon Falls. "With forty-eight hours of mild weather, we could   divert the fire, thus protecting the southeast flank," Poncin says. "But we needed a completed line. An incomplete line posed unacceptable dangers to my crews." The fire might find the gaps in an incomplete line and blow through. Then, instead of the firefighters encircling the fire, the fire would outrun the firefighters.

Yet containment of the blaze continued to prove elusive. As the afternoon burning period of August 21 approached, North Fork took off again. Pushed by a steady southwest wind, its 2-mile-wide fire front hit the cliffs of Secret Valley and split in half.

North Fork was multiplying. Where there had been one fire front, now there were two. One front skirted the Monument Geyser Basin to the north and drove north-by-northeast toward the road between Norris Junction and Mammoth Hot Springs. The second front roared between the narrow walls of Gibbon Canyon then cut northeast for the road between Norris and Canyon Village. Between the two fronts was the Norris Geyser Basin and the historic lodgepole-and-stone Norris Museum.

Once again, North Fork had escaped efforts to contain it. Structure protection at Norris became the top priority as the fire raged on both sides of the development. For two days, North Fork crews fought off the fire at Norris, while Poncin and Area Command quickly upgraded plans to protect Canyon Village.

Beyond the immediate danger to Canyon and the lower-priority threat to West Yellowstone, Poncin had two more problems. The northern head of the blaze was bumping the Norris–Mammoth road at Nymph Lake, above Norris Junction. It seemed likely that the fire would continue running north up the road, on a course straight for Mammoth Hot Springs, the capital of the Yellowstone Nation. And to the south, North Fork continued to press at containment lines 8 miles from Old Faithful.

By August 23, Dave Poncin had come to two major decisions. The first was that the North Fork fire—now stretching over 25 miles with a 100,000-acre fire perimeter—was too big to handle alone. They would have to call in another Type I team and split the fire into two administrative units. One team would handle the south and west ends (Old Faithful, Madison Junction, and West Yellowstone), and the second team would take the north and east ends (Norris Junction, Canyon Village, and the finger heading toward Mammoth Hot Springs).

Size was not the only factor driving Poncin's decision, which he would present to Area Command as a rather urgent "recommendation." Resources and logistics also were concerns. "I had noticed a number of shortages starting to crop up," Poncin explains. "It was harder to get fresh crews. It was increasingly difficult to obtain helicopters, radios, fire-line explosives, and the old fire shelters we were using to fireproof power poles." Splitting off part of the North Fork fire would help Poncin protect his resources. The Western fire season could run deep into October and shortages in August were signs of trouble to come. Furthermore, in logistical terms it had become impossible for one Type I operation to "support" the entire fire. What if the gale-force winds returned and West Yellowstone, Madison Junction, Norris Geyser Basin, Canyon Village, and Old Faithful came under simultaneous attack? "We would have suffered a terrible defeat," Poncin says. He wanted more crews, but virtually every trained wildland firefighter in the country—over 20,000 men and women—already were at work on fire lines all over the West and in Alaska.

Poncin's second major decision was to conduct a large burnout outside of Canyon Village, along the road between the advancing fire and the sprawling two-hundred-structure development of cabins, campsites, lodges, restaurants, and stores. By the morning of August 23, the park's two concessionaires, TW Services and Hamilton Stores, were closing down operations and evacuating over 800 employees and visitors. Two of the park's largest facilities, Grant and Canyon, now were closed. Others would soon follow. Business revenues had dropped by 40 percent and a spokesperson for TW Services announced that the firm would open negotiations with the Park Service to recoup its losses. In July, Bob Barbee had feared not a natural disaster but an economic one. By late August, it looked like he was getting both.

As Poncin's burnout strategy was put into effect—drip torchers would burn away a semicircle of forest around Canyon Village in the hope of diverting the blaze—a series of key Area Command meetings were underway in West Yellowstone. Most of the officials involved felt that the constant meetings were distracting, taking the incident commanders away from their operations posts. But there seemed to be no way to avoid them. As fires mounted across the West, tensions over resource allocations and strategy were increasing between the various command levels in the Yellowstone area and the Northern Rockies region.

Poncin met with the area commanders and Dan Sholly. The second Type I team for North Fork had been approved. The incident commander would be Curt Bates, from the early days of the Clover-Mist fire. But Barbee wanted Poncin on the hot end of the fire, the Canyon Village end, Sholly said. Canyon Village was a major tourism center, worth millions to the park and TW Services. Poncin was the man who saved Grant Village and, moreover, he was a Forest Service man that Park Service officials at Mammoth felt they could trust—a point unspoken at the meeting but later confirmed by Yellowstone officials.

"I argued that I was more familiar with the south end and the needs of Old Faithful and West Yellowstone," Poncin recalls. "Maybe I should stay where I was. Maybe it would make more sense to leave me on the south end and give Bates the north." Sholly thought it over and agreed. The chief ranger knew Bates and respected his firefighting abilities. Events were proving Bates' concern about the Clover-Mist fire spread correct.

Area Command approved the division and told Poncin and Sholly that there would be an important meeting of the Greater Yellowstone Coordinating Committee in two days, on August 26. Area Command, incident commanders, and Park Service and Forest Service supervisory personnel would be present. Poncin and Sholly immediately understood what *that* meant: trouble was mounting; another change was in the wind.

Meanwhile, the new north end of the North Fork fire, an area of some 30,000 acres, from Gibbon Falls to Canyon Village, would be renamed "Wolf Lake." Administratively, North Fork/Wolf Lake would be two fires. In reality, it remained a single fire, and it did not wait for Curt Bates to make the transition in. On August 25, while Poncin was preparing to hand

148   the fire over to Bates, a gusty 15-mile-per-hour wind lifted the
east side of the fire front and threw it at Canyon Village. This
time the burnout worked: Wolf Lake hit it and veered north,
searching for fuel. It halted a mile above Canyon Village, near
the road that ran north over Dunraven Pass and into the
Lamar Valley. Canyon Village was not yet out of danger—the
area was full of unburned fuels, and unstable winds swirled
between the spectacular cliffs of the Grand Canyon of the Yellowstone River, a quarter-mile east of the village—but for the
moment it appeared safe.

Poncin was not as lucky on other sectors of his fire. At
Seven-Mile Bridge near West Yellowstone, the winds of
August 25 caused North Fork to jump containment lines. The
blaze had escaped again. From West Yellowstone it could be
clearly seen burning in the heights near Cougar Creek, 3 miles
north of the Madison Junction road. The only thing between
the town and the fire to the west was 4 miles of heavily
forested country and the thin blue line of the Madison River.
Another wind shift, and it seemed that West Yellowstone was
done for.

Unfortunately, the wind was about to do just that. The
prevailing southwest wind and unpredictable "microclimate"
drafts from the mountains and canyons of the Yellowstone
area had dominated the fires for two months. As the final act
of the drama approached, however, a new player entered the
scene. The weather pattern had begun its shift toward winter
and Canadian cold fronts were starting to move in, bringing
strong winds from the north.

The first signs came on the south end of North Fork, near
Old Faithful. On August 25, the same day the fire was blowing
northeast at Canyon Village and across the Madison Junction
road, aerial observation discovered spot fires over North
Fork's southern line. In other words, the fire was blowing
north on its north end and spotting *south* on its south end.

"We immediately recognized the severity of the situation," Poncin says. "If the spot fires were not extinguished or
contained within new lines, they would pose a direct threat to
Old Faithful." Late in the day, as an inversion settled thick

smoke and cool air over North Fork, the incident commander ordered a crew to hike in to the spot fires and drew up plans for heavy retardant bombing the next day.

At dawn on August 26, Poncin met with Curt Bates at the Madison Junction fire camp and "handed over" the Wolf Lake fire. "Poncin agreed to share his limited resources with me," Bates recalled later, "knowing that I was badly disadvantaged [in the face of a dangerous fire and resource shortages]. That's the mark of a true professional. Other times I haven't fared as well."

Poncin then returned to West Yellowstone to meet with the public and lobby Area Command for more resources. At noon, Area Command convened the meeting of command and supervisory personnel under the auspices of the Greater Yellowstone Coordinating Committee, the bureaucratic arm created years earlier to improve interagency cooperation.

According to several participants at the meeting and a federal review panel, Area Command did not have good news. The national fire scene was deteriorating. Over a dozen major fires were threatening communities throughout the West. They were out of crews. The military was training soldiers for firefighting as fast as possible and BIFC was considering going to regional unemployment offices to hire 4,000 workers. But these were stopgap measures: the troops and workers only could be used in mop-up and support operations.

"Area Command argued that everyone needed to 'face facts,' " recalls a junior Forest Service officer who wishes to continue his career. " 'Yellowstone isn't the only troubled place in the world,' they said." Then the bad news came. "They told us that we wouldn't be getting any more new resources." In fact, Big Mac, the maximum fire gods, had decided that they would be pulling resources *off* the Yellowstone fires. Regardless of any prior agency agreements, specified crews and overhead teams would be sent not to their home units but to the BIFC staging area in Idaho Falls for redeployment. All other specified resources—helicopters, transport vehicles, line equipment, radios, medical units, kitchen units, etc.—would be released directly to BIFC headquarters in Boise.

The room exploded with protests. "I thought they'd gone insane," recalls one Park Service official. "Yellowstone Park was about to burn down! West Yellowstone, Canyon Village, Old Faithful, Mammoth Hot Springs, Silver Gate, and Cooke City were within days of destruction and they wanted to pull resources off! A senior park representative stood up: 'We will release *no crews*,' he said. 'You people can go to hell! We need more crews, more resources, not less!' "

The Forest Service personnel at the meeting were divided. Some still resented Yellowstone Park for letting the fires burn in July and, in their view, continuing to hamper firefighting efforts with an unnecessary emphasis on light-on-the-land tactics and a compulsive zeal for enforcing regulations. In one notorious incident that became an archetype for firefighter attitudes toward the park, a Yellowstone ranger had threatened to ticket a California Division of Forestry crew for taking a truck across a meadow to fight a fire. Many Forest Service firefighters believed the park had a virtual veto on all fire-suppression efforts in the area. Others in the Forest Service—many of them of a younger generation coming up through the ranks—sympathized with the idea of natural regulation and thought Mammoth was doing the best it could in a difficult situation.

The meeting soon calmed down. This was not a request, the Area Commanders reminded the assembly: it was an order. But Area Command needed the cooperation of Mammoth and the incident commanders to effectively implement the order. Not only were the lines of true authority between Area Command, Mammoth, and the national forests still unclear but the incident commanders could delay matters by simply not releasing resources and crews, and by ignoring or "losing" the demobilization orders.

Careers were at stake now. It wasn't just "Yellowstone Park's fires" anymore. If the fires went on through September and the park was reduced to a smoking ruin, heads would roll. Would the blame be pinned on Bob Barbee and the Park Service alone, or would the media drive them all up to some congressional chopping block?

The rumblings of discontent extended to the highest levels of the Forest Service. That day, a "Certified Confidential" memo from a high-ranking Forest Service fire god went out to incident commanders in Yellowstone. "Area Command is taking the flak for hanging on to resources and not putting the fires out," the memo read, although as we've seen Area Command in fact was not fully successful in "hanging on" to resources and the blame remained more targeted at Yellowstone Park. "Area Command is the buffer since you are supposedly doing what they want," the memo continued. "People refuse to acknowledge that the Park is calling the shots within the Park, and continue to expect Area Command to do so and be more responsive to Forest Service views. Stay clear of this."

Stay clear of this. Cover your ass. Be careful. Fight the fire but avoid the bureaucratic war that may decide your fate. Not an easy task. *Things are falling apart at Area Command,* an officer at the meeting noted in his diary.

It was not that the center was not holding: the center had never been clearly established. As the meeting dragged on toward 2 P.M. and the afternoon burning period picked up speed, a new set of guidelines for the fires was hammered out. The basis of the new guidelines was simple: "cut losses."

Cut losses: attempt to hold established lines and concentrate efforts on protecting communities and structures. As for the rest, let it go. They would build no more fire lines deep in the forests. The fires would be allowed to burn anywhere they did not pose a threat to communities or structures. The irony of the situation was not lost on the veteran firefighters. The circle had quickly come around: they were back to a de facto natural regulation.

Dave Poncin, for one, thought he could live with the new guidelines. "I would concentrate my forces around West Yellowstone and Old Faithful. As I drove back to my base camp I remember thinking, 'Maybe things aren't that bad.' Pulling back and cutting loses is a good idea when resources are low. Of course, I was on a high-priority fire. The retardant bombing and ground crews would be getting the spot fires on the south flank near Old Faithful under control by now, and while the fire in the hills outside of West Yellowstone frightened the residents, I believed the town would escape destruction. So things were looking pretty good. I parked my car and walked over to my Operations Chief and Division Supervisor for the south flank. They were looking awfully tight-lipped and grim. What now?"

Area Command had denied the request for retardant bombing of the south flank spot fires, saying it did not fit the new guidelines.

Poncin ran back to his car and returned to Area Command. He knew what was going on. He had been in the fire game a long time. He knew that the Forest Service's Region One coordinator in Missoula, Montana, the authority level above Area Command on the allocation of certain resources, was trying to hang on to a shrinking supply of retardant and active bombers, and had pressured Area Command to start denying those resources. Air tanker priorities were set to protect life and private property when fires taxed resources. An error had been made in overlooking the Old Faithful complex and its proximity to the breached south flank.

Poncin met with the two area commanders and spelled it out on the map. The Canadian cold front was coming down from the north, bringing shifting winds. Loss of the south flank would put an uncontained blaze within 6 miles of the historic Old Faithful development. The two area commanders were sympathetic—they themselves were caught in the bureaucratic crossfire—and agreed to reverse the decision, bucking Region One. Retardant drops, more crews, and a fleet of Chinook helicopters would be on the scene the next morning, August 27. But they had lost a critical day. The spot fires were over the line and consolidating into a mile-wide fire front. It looked like the beginning of the end of Old Faithful— and that wasn't Area Command's only problem. As August came to a close and public pressure for action mounted, big trouble also was brewing at the north end of the Yellowstone fire complex.

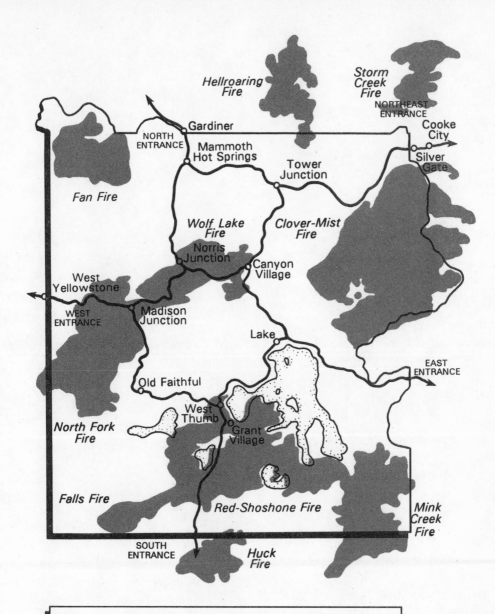

YELLOWSTONE FIRES
September 1

All fires expand. Hellroaring fire has entered park.
North Fork fire threatens Old Faithful, West Yellowstone,
Mammoth Hot Springs, and Canyon Village. Storm Creek
fire threatens Silver Tip Ranch, Silver Gate, and Cooke City.

© Linda Marston, 1993

XIV

Silver Tip

Late in the evening of August 31 a sudden wind shift blew the Storm Creek fire over containment lines and into the Lost Creek drainage, approximately 3 miles north of Yellowstone Park, Silver Gate, and Cooke City. Two crews were cut off by the blaze and retreated to a gravel bar from where, they reported, they could hear the fire "roaring like a freight train" all night long.

Storm Creek had become a priority fire. Four days after Black Saturday, it was turned over to a Type I team led by David Liebersbach, an Alaskan fire god from the Bureau of Land Management. By September 1, there were over 1,000 firefighters and forty pieces of heavy equipment—water tenders, bulldozers, helicopters—assigned to the fire. The arrival of three Army units for backup work allowed additional firefighters to be deployed on newly constructed lines, but it seemed to do little good. The fire, its size now approaching 50,000 acres, had burned out of Lost Creek and into the broad Slough Creek drainage. Directly in its path was the historic Silver Tip Ranch.

Silver Tip was a remote place, built in the 1920s by the

156   Guggenheim family as a vacation hideaway. Joseph B. "Frenchy" Duret had hacked the rough road into the area, and even in 1988 the only way up the trail was by foot, four-wheel-drive or mule train. Frenchy, alas, came to an unfortunate but emblematic end after a life in the wilderness. One day the mountain man met a grizzly bear—also called "silver tip" for the light-colored guard hairs on its shoulder hump—caught in one of his traps. Frenchy put a bullet into the beast. Before he could get off another shot, the enraged animal broke from the trap and attacked him. A terrific struggle ensued. Frenchy's rifle was smashed to pieces. The bear—paw prints found in the area indicated it weighed over 1,000 pounds—clawed and chewed up Frenchy, stripping him naked and leaving him for dead. But the mountain man regained consciousness and started to crawl home. He made it 2 miles before loss of blood killed him. Park rangers found his broken body a day later and buried him on the spot.

Liebersbach had assigned protection of Silver Tip Ranch to Gallatin National Forest Ranger Larry Sears; Frenchy's fate would be a lesson to his crews. Recently a grizzly had tried to break into the building housing the garbage incinerator, leaving behind smashed boards and deep claw scars. The crews, Larry Sears decided, would sleep in the barn, away from the marauding bear. Shirley Blakely, the woman who managed the ranch for its wealthy absentee owner, had another idea. She would put the firefighters up in the woodsily elegant, antique-filled lodge. The thought of his gnarly firefighters amid the splendors of the Silver Tip lodge appalled Sears. The ranger warned them that with one screwup they wouldn't be back in the barn—they'd be out with the bears!

Sears had two Native American crews under his command. One, the Fort Peck Sioux #55 crew, was at a ranger patrol cabin 3 miles to the north, at the top of the Slough Creek drainage. Their job was to protect the cabin and a downed Sikorski helicopter that had been servicing a 300-person spike camp evacuated several days earlier. The other crew, Crow #85, was at Silver Tip.

By early September, over 2,500 Indians were at work on

the fire lines in the West. "As far as I was concerned," Larry
Sears said later, "the Indians are some of the best firefighters
in the country." They worked hard, earning $7.40 an hour
with no overtime pay for long days, bringing home tens of
thousands of dollars to reservations plagued by unemployment and alcoholism. Sears knew it was something more than
the money that brought so many Indians to the fire lines. It
was the discipline too, the frontier craft of it, and the chance,
as one young firefighter told him, "to be somebody."

At the ranch, the firefighters had gone to work. Anticipating a frontal assault by the fire, they had laid hose, set up
pumps, moved wood piles away from ranch structures,
dropped snags, drilled the ranch hands on hose operations
and shelter deployment, and swept the rot pockets and debris
off the old shake roofs. Sears felt the Crow Indian crew had "a
special relationship" to the battle at hand. "This is good country," the Crow crew boss had told Sears as they marched in.
The Great Spirit had put it in exactly the right place.

But when it came to the weather, the Great Spirit seemed to
have deserted them. At Area Command, the fire gods were
predicting the worst. The baffling fires had brought the
nation's top behavioralists back to the scene. On September 3,
as Larry Sears awaited the arrival of the Storm Creek fire,
Area Command was drafting the first in a series of tightly held
Fire Behavior Situation Reports that would be circulated to
incident commanders and other high-ranking personnel.
"What else is left to say?" began Area Command's Situation
Report Number One. "The severe fire behavior has been
described in so many ways. People are almost jaded because
they have seen it all:

1. Super-dry fuels in all sizes/classes that are ready to
    burn at any time of the day or night.
2. Fine fuels that have not recovered moisture with
    increases in relative humidity.
3. Extreme readiness of fuels to torch and crown.
4. High levels of ignition probabilities.

5. Long-distance spotting.
6. Fire growth through repeated cycles of torching and spotting."

The report noted that the fire behavior of the 1988 season was establishing "a critical new point on our curve of fire behavior knowledge" and warned of more wind-driven blazes with the arrival of a dry cold front on September 6 and 7. Finally, it alerted field personnel to the likelihood of two highly significant mergers: the coalescence of the Storm Creek and Hellroaring fires, and the movement of North Fork/Wolf Lake into the west flank of Clover-Mist.

The report did not mention the proximity of the north flank of Clover-Mist to the south flank of Storm Creek— they were about 5 miles apart, with Silver Gate and Cooke City between them. Perhaps the specter of a single gargantuan Clover-Mist/North Fork/Wolf Lake/Storm Creek/Hellroaring fire sweeping like a scythe through Yellowstone Park was too grim to contemplate.

The consolidating fires had "unfavorable consequences to fire-line personnel," the authors of the Situation Report laconically noted. "Fire behavior activity (torching, crowning, spotting, energy release) is accelerated due to the interaction between the fires." Sneak attacks were likely: "Firefighters may be surprised by fire approaching from an unexpected direction."

Silver Tip Ranch was right between the Storm Creek and Hellroaring fires, so for Larry Sears on the smoke-choked morning of September 3 the report meant that he would have to be looking over his shoulder when the inversion dissipated, the afternoon "skylight" opened, and the heat rushed upward, kicking the fire and ground winds into a fury.

Listening to the radio reports of the fire scouts and field personnel, Sears made a critical decision. It would turn out to be his first mistake after more than a week of careful preparations. He was concerned about the Sioux at the Forest Service cabin 3 miles to the north. They had fireproofed the area to protect the cabin and the downed Sikorski, but they were sick and dirty and tired. The crew boss was coming down with bronchitis. Around 9 A.M., Sears set off for the cabin. He had decided to swap the Sioux with one of the Crow squads, allowing the Sioux to get clean clothes, hot showers, and a hot meal at the Silver Tip lodge.

"I thought it would be a big morale booster," Sears recalled. "I had noticed that the inversion didn't break up until late afternoon. I figured I had time to swap the crews and get them back to their posts before the active burning period began." It didn't, however, quite turn out that way.

Telling one of the Crow squads—a twenty-man crew is made up of two ten-man squads—to get their gear together and follow him up the trail, Sears set off through the smoky forest, keeping a wary eye out for bears. He already had radioed ahead to the Sioux, telling them to meet him halfway down the trail. The inversion seemed to be holding things down. "After twenty minutes, I found the fire edge," Sears said. "It was burning spottily with low flames in heavy fuels on both sides of the trail." Sears moved on. The Crow squad came up behind him. Soon they met the Sioux, who were delighted with the news of their brief "vacation" and immediately set off for the ranch.

But something weird was happening. The inversion was breaking up rapidly. At the fire's edge, where an hour earlier there had been only low flame, trees were torching. Sears felt a quickening breeze moving on downstream.

He was in trouble. Switching the crews to give the Sioux team a break meant that some of the personnel preparing for over a week to protect Silver Tip would be at the Forest Service cabin, with the Storm Creek fire moving in between the two sites and down on the ranch.

The forest began to burn hard around them. They broke into a jog, reaching Silver Tip just as the bottom land behind them crowned out in 200-foot flames.

"The meals and showers would have to wait," Sears says. "I turned the Sioux over to the Crow crew boss, who had stayed behind at the ranch with half his team. The wind was

160   picking up. It was obvious that a firestorm was heading our
way. I told the Crow boss he had an hour to brief and drill the
Sioux to take over the positions" of the departed Crow squad.

Sears radioed the Incident Command Post at Cooke City
with an urgent request for air support. Two Bell helicopters
lifted off and within ten minutes were hauling 1,500-pound
buckets of water out of Slough Creek and dropping them in
the trees north of the ranch to slow the fire's approach. Sprinklers were turned on to protect the structures. Sears watched
in dismay as the water hit the parched roofs and evaporated.

By 2:30, the inversion "skylight" had opened and the
downstream breeze turned into a strong gusty wind. The helicopter pilots were reporting extreme turbulence. Sears
radioed up, asking them to move their bucket drops into the
trees adjacent to the structures. The fire, Sears later noted in
his report of the incident, was now within 200 yards and
burning along the north, east, and west sides of the property.
Hot embers were dropping into the compound. Sears clocked
the wind at 40 miles per hour. The roar was deafening.

They were losing it. "I told everyone to assemble in the
burned-out, fenced meadow beside the lodge and prepare for
a voluntary shelter deployment," the ranger says. "We had
practiced it earlier in the week. Forty-eight men and women—
firefighters as well as ranch hands who had volunteered to
stay behind—gathered in the black wind, their faces turned
away from the heat and thundering flames." An enormous
smoke column was building overhead. The wind, increasing
to 60 miles per hour, began to rotate around the meadow in a
cyclonic pattern. Trees snapped in half and were torn from the
ground. Embers the size of tennis balls shot through the air.

The shelter deployment began.

The silver bags bucked and kicked in the wind as the
ranch hands and all but a few firefighters crawled inside.
Then from the barn they heard the screams of terrified horses
and mules.

"I ordered the remaining firefighters to bring three of the
trucks into the meadow and park them around the deployment area," Sears says. "With the barn likely to go up in

flames, we turned out the horses and mules. The animals stampeded into the meadow with us, careening in circles as the fire closed in." The brightly colored trucks in the howling darkness would protect the sheltered figures from the animals' panicked hooves.

Overhead, the helicopters pulled away. The wind was too extreme. The darkness parted for a moment and one of the pilots, as he would later relate to investigators, caught a glimpse of the incredible sight below: a storm-tossed sea of fire surrounding the ranch, the meadow of stampeding horses, and the silver shelters. Larry Sears was crouched beside one of the trucks, counting shelters and trying to figure out if everyone was accounted for, when the pilot of helicopter 664 radioed in a final message, recorded in the official fire-shelter deployment report.

"664 to Sears."

The ranger pulled the radio from his belt and raised his voice over the storm. "This is Sears."

"Partner, you've got yourself one hell of a firestorm down there and there's nothing more I can do for you."

"I understand," Sears said. "Thanks for your help."

The pilot switched to another channel. He hoped Sears wasn't listening. "Incident Command, 664 to Incident Command."

"Incident Command here."

"Stand by for medical evacuation." Pilot 664 figured he would be hauling out bodies.

"Standing by," came the reply from Incident Command.

The firestorm came down around them. A window blew out in the lodge. A horse screamed as a fiery ball crashed into its face. A rain of fire swept the meadow and still the hoses pumped water on the buildings and still a few firefighters stood beside the trucks, awed by the spectacle and ready to fight the hopeless fight. Finally, the wind lessened, the roar turned to a mere rumble. "The storm was diminishing," Sears recalled. "I looked around. The buildings were still there! Our week of preparations had paid off." The hoses and sprinklers had kept pumping, and the fire had not found enough fuel near the structures to torch them up. Spot fires flared around the compound and the crew attacked them with renewed vigor. "We had made it!" Sears says. "We had beat the fire! We did it with our good crews, our training, and a little help from God."

Within an hour, the compound had been secured and Sears allowed the ranch hands out of their shelters. Shirley Blakely, the ranch boss, took her people back to the cook-house and made supper. As evening came, the Storm Creek fire moved south into Yellowstone Park.

At Silver Tip Ranch, everyone sat down in the antique-filled dining room. The lodge was immaculate. "The Crow crew had been caring for it as if it was a museum. Not a plate or lamp was out of place."

Later, Sears adds, "everyone agreed it was the best meal they ever had."

XV

Bozeman

On the morning of September 4, less than twelve hours after the Storm Creek fire swept past the Silver Tip Ranch and into Yellowstone, Bob Barbee put on civilian clothes and drove away from park headquarters at Mammoth Hot Springs. His destination was the city of Bozeman, 90 miles away. With him was Chief of Public Affairs Joan Anzelmo. She too was not wearing her uniform.

The mood in the region was getting uglier as the park skidded from one disaster to another. "The fates seemed to be conspiring against us," Barbee recalled. "Our happy-face message of fire's ecological benefits and the optimistic control assessments of late July had been a perfect setup for a fall." With Black Saturday marking the opening of a stage of extreme fire behavior not seen since 1910, local irritation with the park turned to fury. "I'd never experienced anything like it before," Barbee says.

Barbee's long tenure with the Park Service had not been without controversy. He had been at Yosemite when there were rowdy demonstrations over the future of the park; at Point Reyes National Seashore during a tumultuous dispute

164   with local ranchers; at Hawaii Volcanoes National Park when lava spilled into farmers' fields and burned homes; at Redwoods National Park when Congress voted to seize private timber land in a park expansion project. He was no stranger to controversy. "But this—this!—people were blaming me, *personally*, for the fires!" The affable park superintendent grows visibly upset at the memory. "It was a goddamn lynch mob out there!" Death threats had come into the Mammoth switchboard. Although discounting the seriousness of the threats, Dan Sholly had quietly posted a guard on Barbee's house.

Barbee and Anzelmo drove out through Gardiner and took the long highway up the Paradise Valley, a gray pall of smoke shrouding the mountains around them. It was better to go to Bozeman in civilian clothes. The whole thing was getting out of hand. There was no telling what might happen—the sight of a Park Service uniform might spark a riot.

"Welcome to the Barbee-que," read the signs in the local communities. "Last Stop Before Hell." "Yellowstone National Wasteland."

"The fires themselves were bad enough," Joan Anzelmo recalls, "but the media were making it worse." Anzelmo was desperately trying to put a more positive spin on the message—they *were* fighting the fires; there *had* been successes: at Grant Village, at the Fan fire, on the Thunderer, in stomping out new starts, in the largest assembly of firefighters in recent U.S. history, in preventing fatalities and saving structures. "Yet every bit of bad news seemed to be splashed across the headlines," Anzelmo says, "while the good news was ignored or buried deep in the stories."

The visit of National Park Service Director William Penn Mott was an example of a good opportunity gone bad. On August 30, Mott had visited the park, toured the fires, was briefed by incident commanders, and held a press conference. It was a positive show of high-level concern, with Mott admitting he had not realized the scope of the blazes, but nevertheless defending the natural fire policy. He praised the park and the firefighting effort, telling reporters as Barbee looked on that "you can't blame anybody for what's happened" and adding that "everybody is doing one hell of a job here." In a gesture of reassurance to those concerned about the fires' impact on tourism, Mott announced that the Park Service would launch a major marketing campaign in the winter.

What about the economic damage to the region, a reporter asked. "Some people have lost money," Mott replied, "but others have gained because of the fires." The firefight, Mott concluded, had been a boon to some sectors of the regional economy where over $50 million had been spent.

"Of course," says Bob Barbee, "the media then led the news with Mott's defense of the let-burn policy and his observation that the fires had a positive economic side." It was exactly the sort of thing the region and its representatives in Washington did not want to hear. Congressman Ron Marlenee, Republican of Montana, shot off an angry letter to Interior Secretary Donald Hodel demanding Mott's resignation. The senators from Wyoming, Wallop and Simpson, soon joined Marlenee's call. In a speech on the Senate floor, Simpson declared that "Yellowstone may well have been destroyed by the very people who were assigned to protect it."

The public perception of park officials "destroying" Yellowstone would grow in the coming days and weeks. Jack Troyer, head of the interagency Greater Yellowstone Coordinating Committee, hurriedly pulled together a high-level meeting at the Holiday Inn in Bozeman on September 4 to try to deal with that perception and the increasingly severe fire season. Area Command was there in force. Park Service Regional Director Lorraine Mintzmyer and Forest Service Regional Forester Gary Cargill flew in from Denver. Also attending were the five forest supervisors from lands around the park, two other regional foresters, along with representatives from other Forest Service offices. A new strategy would have to be approved. It was a tense and at times angry meeting. Homes and lives were at stake in Yellowstone; careers were at stake in Bozeman.

Area Command was getting hit from all sides. For the last two days, the fire-projection team had been scrambling to keep up with a rapidly deteriorating weather situation.

166 Rumors were circulating about another dramatic shift in the firefight. The Area Command organization had ballooned to over 200 people. The two Area Command co-commanders— Rick Gale from the Park Service and Ken Dittmer from the Forest Service—were the fourth and final set of chiefs to be rotated through the top of the command structure since its inception on July 23. From below, the incident commanders were pressing them daily for more resources. From above, BIFC and the maximum fire gods were pressing them to release resources. The local communities, congressional delegations, and the media were screaming for action. Due to fires elsewhere, the infrared overflights of the fire complex were being cut back, reducing timely intelligence on the fast-moving blazes. Morale on the fire lines was crumbling.

Despite the fact that the formal delegation of authority limited Gale and Dittmer's mission to setting priorities and coordinating the resource flow, the truth was that Area Command had been drawn deep into tactical control of the Yellowstone fires. Area Command was, in effect, acting as a super incident commander. This position was forced on them by the inherent ambiguity in the relationship between the Area Command level and the multiple incident commands in the Greater Yellowstone Area fire complex. But Area Command still needed the park superintendent and the forest supervisors to sign off on major strategic changes—a need that further confused the situation. The two co-commanders also needed political cover within their respective bureaucracies. If the gateway communities or Old Faithful or Canyon Village burned up, some of the legion of captains steering the overloaded ship of fire management were sure to get thrown overboard.

Anzelmo and Troyer set up a video camera in the back of the room to record the event. The four-hour videotape of the critical meeting reveals the deep management divisions hidden from the public during the summer. By the end of the meeting, there would be an announcement of startling new developments concerning the fate of Old Faithful and the gateway communities, as well as a new strategy for coping with the blazes.

Following a brief introduction by Mintzmyer, Barbee walked to the front of the room to address his colleagues. He had a few things to get off his chest. Yellowstone Park, he said, "has been experiencing absolute and utter frustration. The situation has degenerated from cheap shots to a piling on of accusations to what is now a virtual lynch-mob atmosphere. The Park Service, and especially Yellowstone, has been made the villain. We're responsible for everything that has occurred. That's the general public perception: that the National Park Service fire-management policy and its thoughtless application is responsible for absolutely everything that's gone wrong. Now, that's bullshit—and I know you all know that—but nevertheless that's what's out there."

The Forest Service was allowing and even encouraging the public view that the Park Service was to blame. "For instance," Barbee said, angrily ruffling a sheaf of notes and news clips, "you have a Forest Service fire-management officer making a public statement that the North Fork fire could have been stopped if only the Forest Service had been allowed to pursue it into the park. That news went out like wildfire. It caused a hemorrhage that we are dealing with every day. I could give you other examples, but you all know what I'm talking about.

"There is absolutely no interagency identity to this whole enterprise," Barbee bitterly added. "I don't see it. I don't feel it. I realize that the interagency fire teams are largely dominated by U.S. Forest Service personnel, and they do an exemplary job, but there is no interagency identity. We go to public meetings and the Forest Service is applauded, while we are the villains. A few weeks ago, our sense of unity and direction and the spirit of cooperation was generally high. That situation does not exist today. What happened?"

Barbee suddenly was looking at the likelihood that his long and successful career in the Park Service would abruptly end in disgrace. "These fires are so complicated!" Barbee exclaimed. "The competition for allocation of resources has become such a rivalry that there is a lot of raggedness out there and a lot of nasty little situations are developing—no one wants to surrender resources. And frankly, *I'm tired of it.* My staff is tired of it. They're near revolt right now, and I'm sure I speak for the entire Yellowstone organization. I'm looking for this meeting to turn things around. We've either got to pull our chestnuts together or this whole damn thing is going to fall apart!"

Mintzmyer, Barbee's superior, tried to smooth things over. "What we need to do now," she said, "is to make the image of this group a positive one. We need to be working together to make something happen, to get on with the mending process and the directional process. First, though, we need to know exactly what the current situation is." She turned the floor over to Rick Gale, the Park Service co-commander of Area Command.

Gale was a tall man with an authoritative military presence. Like Ken Dittmer, his Forest Service co-commander, he had been around the fire game a very long time. He got down to business right away.

"The Yellowstone fires were at 933,000 acres as of this morning," he said. "We expect them to go over the million-acre mark by tomorrow. We've built over 350 miles of line. Of that, we've held only 29 miles. We've dropped 700,000 gallons of water and 1,250,000 gallons of retardant. At present, we have 9,600 people working under an Area Command that covers 100-by-90 air miles, that's over 10 million acres. We've spent over $67 million. There are currently eleven incident-management teams in the area, six of which are Type I teams. Three communities are under immediate structure protection—the Island Park residential area 18 miles southwest of West Yellowstone, West Yellowstone itself, and the Cooke City/Silver Gate area. Our resources are dwindling. It's becoming a question of which house do you burn in order to be able to move the engines to another house."

He paused, taking a deep breath. The Yellowstone fires had moved off the charts. "We are at a complexity level that exceeds anything we have known before," Gale said. "We can-not do anymore standard firefighting. We cannot build line and hold it in the traditional ways. We're at a position in time and resources and stress where some very hard decisions will have to be made."

Ken Dittmer joined Gale at the front of the room. Follow-ing a summary of the position of each fire, Dittmer tried to ease into the bad news. There was a lot of it.

"We're looking," Dittmer said, "at a significant reduction of manpower due to attrition of crews and the persistent demand for resources throughout the West. We've already released fifty-two Type II crews—over 1,000 firefighters—and seven helicopters. We're going to have to demobilize more crews in the coming days."

A muttering of dissent swept the room.

"Our people are exhausted," Dittmer said. "They don't want to give in. They're professional wildland firefighters and they've been beat over the head and beat over the head and beat over the head—they wanted a chance to beat back. But they never got that chance," Dittmer sighed. "And that's really the best way to depict what's happening out there on the fire line—you have 9,600 people that are gettin' worn out and frus-trated.

"So it's time to focus on areas we can't afford to lose," he continued. "We're going to pull back from the fire lines and put our marbles in the protection of Cooke City, Silver Gate, West Yellowstone, Island Park, Old Faithful, and Canyon Vil-lage. As for the rest, we'll deal with it when the weather decides what it's going to do."

Barbee interrupted from the audience. "This decision is a result of your meetings with the incident commanders, right? I mean, it's not just you two guys coming up with this?"

Dittmer glanced at Gale. Some people in the room still weren't getting it. "I'm not going to sell you a bill of goods and say that the individual incident-management teams are in agreement with doing this," Dittmer replied. "Some of them have been fighting this for seventy days and they don't want to quit. Put yourself in their shoes. You have eleven incident commanders out there. Each one says, 'Well, if we could just have another ten crews, a dozer, a helicopter, maybe we could put a line here, a backfire there, fall back to this point and try again.' They don't want to quit! But we're saying that given the weather and fuel conditions, we could have five times as many resources and our chances of success wouldn't be any greater. We'd just spend a bunch more money and put a lot of folks in jeopardy."

The implications began to sink in. The fires would run free except around the communities. The "resource mix" would change: more crews and helicopters would be released; more engines and dozers would be brought in to protect structures. Area Command could not pull off such a change without the approval of the officials in the room. "I don't envy you having to make these decisions," Rick Gale said.

"We just can't fight everything," added Ken Dittmer. "And we've got to realize that nothing is stopping these fires. Roads don't stop it. Rivers don't stop it. Wet meadows don't stop it. Young lodgepole growths with no ground cover don't stop it. Topographic features don't stop it. *We cannot hold fire line.*"

In terms of the fire strategy, a gigantic leap was taking place. It was the second major shift in ten days. Nine days earlier, at the West Yellowstone Area Command meeting of August 26, shrinking resources had led to the decision to create no new fire lines, while attempting to hold existing lines.

Now, established lines would be abandoned and the manning of some fires would be reduced so low that the effect would be to turn the blazes loose. Resources would be pulled off almost all the newly designated "interior lines"— that is, virtually everything inside the park, with the exception of structure protection at Old Faithful and Canyon Village. A new "exterior line" would trace south from Silver Gate and Cooke City, and down along the Clarks Fork Valley and the Crandall Creek area of the Sunlight Basin. A parallel effort would be conducted 60 miles away, around West Yellowstone and Island Park. As for the rest—as for Hellroaring and most of Storm Creek, as for the south flank of Clover-

Mist, the multiple northern heads of North Fork, most of the Snake River Complex, and most of Mink and Huck—they would let it burn.

There was bitter opposition to the plan, but Area Command believed it had no other choice. Resources weren't the only problem. Another powerful Canadian cold front was on the way, Dittmer told the Sunday meeting, bringing the worst possible weather pattern, a deadly dance of shifting winds.

On Monday and Tuesday, the front would send 20 mile-per-hour winds down from the north and northeast, pushing the west flank of the North Fork fire below West Yellowstone and right at Island Park, and expanding North Fork's south flank near Old Faithful. Simultaneously, the Storm Creek fire would be pushed into the canyon cut by Soda Butte Creek west of Silver Gate and Cooke City. "It's a narrow canyon there," Dittmer warned, "with some of the heaviest fuels in the entire complex."

By the end of Tuesday, the fires would be "primed, cocked, loaded, and aimed," Dittmer said—right at West Yellowstone, Old Faithful, Silver Gate, and Cooke City. Then, when the full force of the cold front hit around Wednesday, the winds would shift, blowing in from the southwest and increasing to more than 30 miles per hour. The high southwest winds would blast the fires straight at the communities. In the conference room, reality sank in. A gasp went up from the audience.

"We're increasing structure protection at all sites," Dittmer said, "and we've evacuated Silver Gate and Cooke. We're doing a last-ditch burnout between Silver Gate and the Northeast Entrance of the park, to try to have *something* to prevent a flaming front from Storm Creek coming right into the community. But I have to tell you, we give Silver Gate and Cooke City less than a 20 percent chance of survival."

XVI

Storm Creek

Long before dawn on Tuesday, September 6, the Storm Creek base camp east of Cooke City began to clank and grind to life. Bright white lights cut the darkness. Sleepy firefighters shuffled over to the mess trucks for sausages, bacon, eggs, hotcakes, fruit, and coffee, lots of coffee. Generators, transport vehicles, water tenders, and the nearby helipad groaned and rumbled into another day.

Incident Commander David Liebersbach had gathered his top firefighters around a big map tacked to a bulletin board. A few reporters were with the group. The Yellowstone fires had passed the million-acre mark, and news of the imminent demise of Silver Gate and Cooke City had spread across the nation.

The overhead officers and crew bosses waited for the IC to speak. Some of the line firefighters, the grunts, the "ground pounders," rose from the long breakfast tables and drifted over to the group, balancing steaming coffee cups on plates of rapidly cooling food.

The burnout between Silver Gate and the northeast entrance of the park was not going well, Leibersbach said.

174   They had been at it for over thirty-six hours. The dozers had cut a 100-foot-wide, 3-mile-long strip through the forest. By 3:30 P.M. on Sunday, September 4, while the fire gods were meeting in Bozeman, the aerial firing ship had been overhead, dropping ping-pong balls into the forest. A half-mile to the west, at the Northeast Entrance, the Park Service had ignited a burnout around the ranger station. If they managed to burn out the zone between the dozer line and the Northeast Entrance, Silver Gate and Cooke would be provided with some protection when the Storm Creek fire dropped into the narrow Soda Butte drainage; with a bit of luck, the burnout fire would burn into the main fire.

More firefighters joined the crowd. They silently listened as a gray dawn sharp with the smell of burned wood broke behind the mountains. The projected wind shift, Liebersbach said, had come earlier than expected. They had spent most of yesterday chasing spot fires from a mild southwest breeze, the wind pushing embers from the burnout in precisely the wrong direction, over the dozer line and toward Silver Gate and Cooke City. Heavy smoke was hampering reconnaissance and helicopter bucket drops on the fires. The Storm Creek fire itself was hanging somewhere back in upper Pebble Creek inside the park, two miles to the north.

The smoke, the steep slopes of the Soda Butte drainage, and the shifting winds were making the burnout effort a test of wills. High winds from the west and southwest were due in soon, Liebersbach said. The winds and low visibility made air support extremely unlikely. The dozers were cutting lines around the towns. There was a slight chance—if the winds were more southwesterly than westerly—that the fire would hit the partial burnout and veer into the north hills, where a heavily manned dozer line might keep it from coming down on the towns. But the odds were that sometime today or tonight a firestorm would come blasting like a blowtorch up the Soda Butte Canyon, destroying Silver Gate and moving on to Cooke City.

"You all know your duties and you know where the safety zones are," the gruff Alaskan said, according to eyewitnesses.

Dawn showed the exhaustion and worry in their faces. "I know you'll do your best. You've been fighting hard under tough circumstances. These are historic fires in a historic year of fire, and we've been getting a lot of attention and a lot of pressure, but I want you to know my priorities remain the same. My first priority is safeguarding your lives. My second priority is protecting the structures in Silver Gate and Cooke. If we have to let 'em go, we'll let 'em go. If we have to pull out, we'll pull out and fight another day. But no matter what you do or what the fire does, every one of you are gonna be heroes, and by God every one of you will be alive when we get done with it."

The crowd broke up into a series of smaller meetings. There was much to do: the base camp itself might be overrun by the fire, so every detail of the contingency evacuation plan had to be gone over; despite Liebersbach's request that county officials implement a forced evacuation of the towns, some residents had remained behind—they had to be briefed on safety zones and last-minute evacuation procedures; the media had to be dealt with; Area Command had to be updated; the military crews assisting the effort had to be given safety notes and assignments; and the battle plan for the day had to be reviewed in light of the latest intelligence and weather forecasts.

Around noon, Liebersbach climbed into a jeep and headed for the dozer line. Entering Cooke City, Liebersbach noticed that a local wag had added a "d" to the sign marking the town, changing it to "Cooked City." At the Second Edition Cafe, Joan and Bill Humiston, lifelong residents of the area, had finally pulled out, leaving behind a final message on the board that usually advertised hearty dinners and famous homemade pie: "God Be With Us."

Trucks, hoses, and firefighters lined the street. The big dozers were working the perimeter of the hamlet. Area Command had managed to shift more resources to the site: Liebersbach now had 1,200 firefighters, 27 engines, 6 bulldozers, 7 helicopters, and 5 C-130 air tankers at his command.

* * *

176   Still, problems and confusion persisted. By noon it had become obvious that the heavy smoke would not allow the air tankers into the narrow mountain-flanked drainage. Maybe if Liebersbach had gotten the resources he wanted, he wouldn't have needed the air tankers.

It later emerged that Liebersbach had wanted the more powerful helitorches, not ping-pong balls, for the burnout. His operations staff believed the torches would give them a cleaner and wider burn. He was told the helitorches were not available.

There also was the problem of structure protection. Structure protection of Silver Gate and Cooke City was not fully under the authority of the Storm Creek commander. Because the initial threat to the towns had come from the *Clover-Mist* fire, on Black Saturday and after, structure protection had remained under the authority of the Clover-Mist commander, whose base camp was 20 miles away at Crandall Creek.

A fitful attempt had been made to create a "Unified Command" of the Storm Creek and Clover-Mist teams in the Cooke area, inserting yet another level of bureaucracy into an already confused situation. The idea was that the Storm Creek crews would handle the wildfire and the Clover-Mist crews would take care of the structure protection. But this command was "unified" in name only. Because of the overlapping command structures, the right hand of the fire-control effort often did not know what the left hand was doing. At one point, for example, according to the official review of the fire, the Storm Creek command was running a burnout on a ridge above Cooke City, while the Clover-Mist command was burning out along a road directly below the ridge, a potentially deadly violation of basic safety procedures.

In Silver Gate, 3 miles down the road from Cooke City, departing residents had hung American flags from the balconies and windows of their rustic wood homes, in a final salute to the firefighters. Several hundred firefighters were deployed around the town. At the town's west border—the side closest

to the park and the burnout—Grizzly Lodge owner Hayes **177**
Kirby had run the Texas state flag up the pole below the Stars
and Stripes.

Kirby had bitterly fought Liebersbach's burnout plan,
arguing instead for a massive frontal assault on the fire, but
that fight was over now. The Air Force veteran had changed
into fire-retardant clothing and was waiting for the end. "I
wasn't suicidal," Kirby recalled, "but I was going to defend my
home and be on the last truck out of town." One side of the
sign outside the Grizzly Lodge read: "Give It Hell, Guys. It
Ain't Over Till It's Over." The other side held a shorter message: "Thanks, Mr. Barbee."

Across the narrow road, close by a barricade, a fire-infor-
mation officer stood on the steps of the Range Rider Inn,
describing the dimensions of a firestorm and the proper use
of a fire shelter to a small group of unusually quiet reporters.
A firefighter pulled aside the barricade and let the incident
commander through. He drove a half-mile down the road to
the dozer line. The decision to put the line in and burn out the
forest had come three days earlier at a stormy meeting inside
Yellowstone Park's Lamar Ranger Station. Dan Sholly and his
deputy, Steve Frye, were there, as were other park firefighters,
the incident commanders for the Storm Creek, Hellroaring,
Wolf Lake, Clover-Mist fires, Area Commander Ken Dittmer,
fire-behavior experts from Area Command, and representatives of the Gallatin and Shoshone forests. The park was sensitive to local charges that they were secretly hoping the fires
would burn down the communities, thereby allowing Yellowstone to make an expansionist land grab.

With Storm Creek now inside the park and threatening to
move out to the gateway communities, fire lines outside the
park, Mammoth officials knew, would only reinforce the
notion of evil deeds emanating from the Yellowstone Nation.
But the only other place to mount a burnout was in the meadows near Pebble Creek, and that meant heavy dozer work
inside the park. The fire gods, according to the official
account of the meeting, argued that a Pebble Creek burnout
would not work. "It wasn't the dozer use inside the park that

bothered me," Liebersbach recalled, "but the fact that there was not enough time to get resources into Pebble Creek and still be able to pull back if we needed to protect Silver Gate and Cooke City." A line near Silver Gate and a burnout to the Northeast Entrance was the only hope.

By the afternoon of September 6, that hope was fast disappearing. The wind was steadily increasing. By 5 P.M., it was up to 25 miles per hour and more spot fires were appearing. Liebersbach ordered a retreat from the line.

The spot fires were working their way toward Silver Gate. In town, the structure-protection teams began draping blankets of foam and water over the buildings. As darkness fell the cosmic rumbling came closer and soon the firefighters saw flames jumping hundreds of feet into the night sky. Below the flames, bright red smoke boiled out over the black hills.

A reporter on the scene remembers an eerie pause right before the fire hit, a sudden silence, a snowfall of ash, the firefighters looking at one another with expressions of anticipation, excitement, and terror. It was the moment they lived for and dreaded, the moment that defined their existence and might define them out of existence.

Then, suddenly, it was there in the hills behind the Range Rider Inn. Deep in the woods an ember had landed, a thin trail of smoke traced up from the forest. A spot fire. Someone pointed. There! A gust of hot wind tossed the treetops, the air crackled, and a column of flame leaped up the tree, exploding into its crown, tendrils of fire tossing angrily. Another blast of wind hit the tree and threw a wave of embers into the forest and then the roaring flames and crashing trees and shouting voices were everywhere. The spot fire quickly grew to 5 acres. The great pines exploded, each explosion throwing long arms of sparks across the dry hills. Within ten minutes, the hillside north of Silver Gate was a gleaming mass of bright orange rods and red smoke.

But with the southwest wind at its back, the fire had skated north of Silver Gate. No structures were lost. It was an ambiguous victory for the Yellowstone firefighters. They were fighting the unsuccessful burnout, after all, not the Storm Creek fire. The fire itself continued to burn northwest of the town, in upper Pebble Creek, inside the park. The hills north of Silver Gate were forested no more—they had become a graveyard of black posts and glowing embers. And the escaped burnout was moving on, right at Cooke City, 3 miles away.

In Cooke, firefighters readied for the assault. The wind was up to 30 miles per hour, with gusts up to 70 expected, far exceeding the estimates of the weather forecasters.

Around midnight high winds began bending the trees and flakes of ash raced out of the glowing night sky. A great army of wind-driven flame was marching east from Silver Gate. The hills north of town burst into towering flames. Night was transformed to fiery day and in the red glare a man alone in the suddenly deserted street looked up to see the firefighters strung out along the dozer lines north of town, tiny figures below a mountain of flame. Then balls of fire started exploding from the hills above Cooke Pass, outside of town by the Storm Creek base camp. At the camp, the military and the helicopters had been evacuated hours earlier. Trucks were rumbling away. The fire was hanging north of the road—they still could make it out of Cooke Pass and fight another day. Men hurriedly gathered equipment together. They would fall back to the Clover-Mist base camp at Crandall Creek.

But crews were missing, the fire was raging, and one turn of the wind would bring the blaze down on Cooke City. Liebersbach got a radio relay through to Area Command. They were abandoning the camp, he reported. They needed reinforcements. They needed support to regroup fast at Crandall Creek and hit the fire hard come dawn.

But come the dawn of September 7 there would be little help from Area Command. Storm Creek was not the only fire running wild. North Fork was on the move too—thundering down off the Madison Plateau and sending a wall of fire straight for the crown jewel of Yellowstone National Park.

XVII

Old Faithful

The North Fork fire had been on a steady course for Old Faithful since the loss of the southern line on August 26. On August 31, Dave Poncin's team had transitioned off the blaze, turning it over to a Type I team led by California firefighter Denny Bungarz. Poncin and his team left the North Fork fire exhausted, sick from smoke inhalation, and frustrated by the elusive blaze. "Yet," Poncin would later recall, "we also were quietly pleased to have been a part of the great fires. They were endurance fires—testing us, teaching us, and reminding us to be a little humble before them."

North Fork now covered 100,000 acres and showed no sign of stopping. It continued to advance on many fronts. By September 6, as the Storm Creek burnout raged above Silver Gate and Cooke City, fingers from the northern flank of North Fork were reaching out for Mammoth Hot Springs. The northwest flank of the fire had advanced to within a mile of West Yellowstone and was threatening the residential community of Island Park, Idaho. The shifting winds had driven the eastern sectors of the fire—administratively now the Wolf

182   Lake fire under a separate Incident Command, but in reality a major flank of North Fork—4 miles north and 3 miles south of Canyon Village. Wolf Lake Incident Commander Curt Bates, still worried about runs at Canyon Village, feared that the high winds forecast for September 7 would "massively expand" his fire—blowing it north into the Lamar Valley, where it might merge with the Hellroaring fire. Alternatively, Bates worried, a wind shift from a southwesterly to a westerly course could throw Wolf Lake right into the Clover-Mist fire. Either scenario would create a fire zone that would cut Yellowstone Park in half.

At the south end of North Fork, Denny Bungarz worried about the big wind too. His main concern on September 7 was protecting the Old Faithful complex, particularly the historic Old Faithful Inn. "Loss of the Inn," recalls Bob Barbee, "would be a real tragedy. It's not hyperbole to call the seven-story building one of the architectural wonders of the world."

The Inn was built in 1904, entirely from lodgepole pine and locally quarried rhyolite. Its old central section was dominated by a steep shake roof 80 feet high and 350 feet long. Inside the grand central section were 140 guest rooms, two levels of balconies circling the lobby, dining areas, offices, alcoves, and sitting rooms, all of pine. A giant stone fireplace, with eight hearths on the floor and balconies, rose over 70 feet from the lobby and into shadowy beams and rafters. The rich, dark interior was filled with the romance of the Old West— Yellowstone Park and the frontier spirit incarnate. Two later additions did nothing to diminish the romance of the place, and brought the guest rooms to 300 and the total length of the Inn to 800 feet. A beautiful pine porte cochere projected out over a driveway and everywhere was a multitude of woodsy embellishments: dormer windows set with crossed poles, massive pine columns and corner posts, diamond-paned glass, elegant balusters and staircases. Crowning the old central structure was a rooftop observation platform, flying the flags of the United States and the State of Wyoming, and pennants representing the National Park Service and the Yellowstone concessionaire. Long closed to the public, the platform would provide the view of a lifetime when the North Fork fire came roaring in.

But there was no need to be on top of the Old Faithful Inn in order to see the North Fork fire. Three miles away, two enormous columns of smoke were boiling up from behind a forested ridge. One column was due west; the other was southwest. Overhead, the sunny sky was perfectly blue—a mocking reminder of ordinary days in Wonderland.

Bungarz had ordered the evacuation of over 1,200 tourists and workers, yet at noon visitors were still milling about in the big parking lots, waiting for buses. The media had assembled for the deathwatch and Bungarz was quietly assuring them that the area was "defendable."

A menacing, surreal air closed over the geyser basin. The sky was blue, the geysers sprayed diamond washes of water, and a placid herd of buffalo moved across a stream-laced meadow. But to the west two great columns of smoke climbed toward heaven. Helicopters circled the complex. Air tankers, tiny as insects before a storm front, traced along the far ridges, dropping little patches of pink, like bits of cloud.

Among the fire watchers were Paul and Suzanne Strasser, founders of the nonprofit Geyser Observation and Study Association, an organization dedicated to studying geysers and hot springs. Paul Strasser recalls that a light rain of ash began to fall around 3 P.M. The wind was picking up from the southwest. "Then we saw the flames," Strasser says. "The forest began to torch out. The black smoke was leaping from the burning trees and melting up into a giant cream-colored cloud, a mushrooming mass streaked with orange and pale green gases."

The air tankers and helicopters kept on coming, now dropping their loads at the perimeter of the complex. Firefighters scurried over the Inn, switching on systems that sent a broad river of water coursing down the wide shake roofs. Atop the observation platform, a number of small figures watched the blaze approach. The Wyoming state flag and the pennants had vanished, but the American flag flew stiff in the strong wind and darkening sky, pointing northeast.

"The southwest firestorm—that southwest column of smoke—was the one that was going to do us in," Strasser says. "The fire was down in the Iron Spring Creek drainage and the smoke column, leaning out in front of the blaze, was unfolding like a speeded-up time-lapse movie of a thunderstorm, with a sharply defined, mushrooming head coming over us. Then a breeze hit us from behind. The fire had begun to suck air into the flame front. The temperature rose sharply and embers started to fly across the parking lot." It was time to get out. The Strassers picked their way carefully across the parking lost, ducking their head against the flying embers.

The smoke column covered the sun and the world was bathed in a dull orange light, an all-encompassing dusk. Then the high winds hit: a howling blast of dirt, ash, embers, and branches, gusting to over 50 miles per hour. "I switched on the engine," Strasser says, "wondering what kind of heat it took to make a car explode." They went north along a service road and kept driving—as far away from the fire as possible.

It was 3:30 P.M. A half mile to the east, rangers Lee Whittlesey and Gary Youngblood were manning the roadblock while a fire crew from Texas worked nearby. Whittlesey could see the Inn from his post. "I was monitoring the radio traffic and had learned that my girlfriend, Terri, a park ranger, was on the observation platform with Denny Bungarz and park scientist Rick Hutchinson," Whittlesey recalls. "The first sign of real trouble came when I heard a message from the observation platform: 'Uh-oh,' a voice said. 'I think somebody better come look at this.'"

Whittlesey slipped the radio back onto his belt. "From close by came the sound of a freight train. A hot wind washed over us. I looked across the road and saw a huge wall of flame 50 yards away in the forest. It was coming right at us and stretched off a mile in both directions."

The appearance of the flames, Whittlesey adds, "made an immediate impression on the Texas fire team. 'Texas crew! Texas crew!' the fire boss shouted. 'Run! Run! Every man for himself! *Run!*'" And then they all were running, the Texas crew, Whittlesey, Youngblood, running for the Old Faithful complex and the only hope of safety—the parking lot.

"I ripped the radio from my belt and shouted, 'I'm abandoning the barricade! I'm abandoning the barricade! The fire's here and it's huge!' " Whittlesey says. The orange air turned bright crimson, brighter, brighter, then went black as night. They were inside the smoke column. The Inn disappeared from view. The wind screamed in their ears and embers the size of fists punched across the road, trailing long arms of wicked shimmering fire. "I remember thinking, *So this is how it ends,*" Whittlesey says, *"this is how I die."*

Weak headlights came into view. A blast of wind battered them. The headlights drew closer. "We had made it to the parking lot! I hit the ground and started crawling. I needed air, down there below the smoke."

A huge fireball roared overhead, streaking across the lot and hitting Observation Point on the far side of the complex. They were surrounded by flames. Nearby, a cabin exploded, engulfed in a rolling wave of fire. Twenty-six structures were lost that day.

Whittlesey and Youngblood stumbled on. They ran into Amy Vanderbilt, a public-affairs official who had been married at the Mammoth chapel five weeks earlier. "She looked terrified," Whittlesey remembers. "I told her, 'We gotta get out of here.' Her husband, Gary Moses, was fighting fire somewhere on the compound; so was my girfriend—if they still were alive."

Through the blackness, Whittlesey saw men fighting a spot fire on the roof of the Inn. "It looked like the Inn was going up."

Youngblood ran off to find a vehicle. Whittlesey looked around, making out other figures in the black wind. A television news crew was standing next to him. They leaned into the wind, the camera pointing at a burning cabin.

The film from that episode has survived and part of it appears in the documentary, *Inferno at Yellowstone.* The reporter shouts into his microphone: "There are extreme

186 winds. Fire is flying all over the place!" Flames cascade over a microwave transmission tower, engulfing it. A gust knocks the cameraman over. "Extreme heat!" the reporter shouts. The cameraman struggled to his knees. "We're going through low visibility to . . . ah . . . extreme heat . . . can barely breathe. . . . The buildings are on fire! The buildings are on fire! The woods on the other side of the geyser are burning! The winds are now unbelievable!"

"Unbelievable was right!," says Lee Whittlesey. "I was astonished. Those lunatics were actually enjoying it!"

Just then Gary Youngblood screeched up in a car. " 'Get in!' he shouted, opening the door. I climbed in." Amy Vanderbilt had disappeared. Other reporters joined the television crew.

"We decided to patrol the perimeter," Whittlesey says. "There were plenty of folks covering the Inn. We'd only be in the way."

The reporters were setting off across the parking lot. Whittlesey rolled down the window. " 'Hey!' I yelled. 'Where do you think you're going!' "

They pointed to Observation Point, where the fireball had landed. "I told them not to go," Whittlesey says. "They would endanger themselves—worse, they would endanger firefighters who might have to go in to rescue them. But I knew what they were after. Observation Point would provide them with a panoramic picture of the burning basin and Inn." The reporters ran off into the swirling smoke.

"Youngblood and I looked at each other," Whittlesey recalls. "We hadn't been feeling too good about the press lately. Well, let the bastards die. They had been warned. We had work to do."

There would be plenty of work. The big wind swept across all the vast Yellowstone country on September 7, the most voracious day of fire after Black Saturday. North Fork grew by 50,000 acres. A ring of fire now virtually surrounded Old Faithful; from its east side, a dagger of wind-driven flame thrust 3 miles out onto the lodgepole stands of Yellowstone's Central Plateau. West Yellowstone continued to be menaced by the fire's northwest flank. At Canyon Village, Curt Bates was battling the Wolf Lake portion of the North Fork fire at the doorstep of the visitor's center. North of Canyon Village, the wind was pushing the hydra-headed fire deep into the Washburn Range, pushing it toward Tower Junction and the road west from Tower to Mammoth. Clover-Mist, Snake River, Hellroaring also were on the move. The empire of fire, and the armies of the wind, now ruled the Yellowstone Nation.

XVIII

Clover-Mist

The big wind swept on from Old Faithful, blowing east at speeds up to 70 miles per hour. It blew across the forested Central Plateau scattering spot fires, across the Hayden Valley flattening the high dry grass, and up onto the Mirror Plateau, where it slammed into the cold west flank of the Clover-Mist fire. Beyond the Mirror Plateau, a great barrier rose in front of the wind—the 9,000-foot ridges of the Absaroka Range, the formidable natural barricade that was to have provided Clover-Mist's ultimate confinement. Beyond the range lay Silver Gate and Cooke City and the ranches of the Clarks Fork Valley and the Clover-Mist base camp at Crandall Creek, remote outposts of paradise, places the fire was never supposed to be.

But there it was. Clover-Mist fire perimeters were now approaching 300,000 acres. Over 1,500 firefighters were assigned to the blaze under Incident Commander Larry Boggs. Into every paradise comes a snake, and this snake was a screaming wind, literally screaming—a hot howling whistling crashing cracking wind hurtling off the Mirror Plateau and hitting the Absaroka Range, shearing north and south, searching out the narrow passes and stream drainages, hungry for fire.

High in the Absaroka Range, the wind found the leading edge of the Clover-Mist fire. It drove the fire between mountain walls thick with old lodgepole and spruce. It funneled the fire into the drainages and passes, concentrating its force into tunnels of flame and wind that in a combustant instant blew down towering pines and consumed them in a superheated vortex.

The fire burst eastward, out of the mountains. The south end of Clover-Mist exploded anew, running over 12 miles in Jones Creek in a single day. Below Jones Creek, Yellowstone's east entrance road to Cody and Buffalo Bill's historic hunting lodge at Pahaska Teepee were threatened. The area was hastily evacuated. Military personnel hacked a fire line from the Cody road to Jones Pass.

Twenty-five miles north of Jones Creek, across blackened and burning forests, Clover-Mist breached the dozer lines Larry Boggs had put in at the creek drainages above the Clarks Fork Valley. The big wind had made it through the Absaroka Range and was pushing the blaze down the forested slopes of the valley. Here was a new land for fire to conquer. At the foot of the mountain slopes, the firefighters were waiting.

Smoke columns climbed up from behind the Absarokas. Directly above the valley, however, the sky was a thin bright blue, dappled with clouds. At the Few Acres development—a cluster of 21 cabins along the Clarks Fork River—a Yellowstone Park crew led by Ranger Bob Mahn was hurriedly completing the structure protection. Brush had been cleared, hose laid, pumps primed and ready to douse the buildings.

Mahn and David Anderson—the leader of one of the two squads that comprised Mahn's crew—had been fighting Clover-Mist nonstop for over three weeks. They had been near Soda Butte Creek on August 16, with Phil Perkins on the Thunderer on Black Saturday, on a Mirror Plateau fire line near Amethyst Mountain on August 26, and at a dozer line in a creek drainage near the top of the Clarks Fork Valley until a midnight call on September 6 alerted them to get ready to go to Few Acres.

Illness and exhaustion had reduced the crew to seventeen firefighters. Still, by 3:45 on the afternoon of September 7, as North Fork raged around Old Faithful and Wolf Lake battered Canyon Village, Mahn's crew had managed to put a basic structure-protection plan for Few Acres into place. The fire would give them little time to rest.

At 4 P.M., remembers squad boss David Anderson, "The wind was blowing like mad. Suddenly it quit. A strange eerie quiet permeated the thick smoke. Nothing stirred. Then we realized that the heat of the approaching fire front was so strong that it was blocking the wind and we were directly in the firestorm's path. It turned dark as dusk except for an ominous glow to the west. Flakes of ash and burned needles began raining down, then firebrands began to fall. Suddenly the fire was sucking in the surface air and at the same time we heard the pulsing roar of the approaching front. Intense heat hit us as the wind switched. Trees torched and clouds of billowing smoke ignited."

The firefighters poured water on the structures. Spot fires broke out in the compound. The fire leaped and danced around them, racing east with the big wind from Old Faithful at its back.

The fire blew right over the Clarks Fork River, jumped the highway and kept rolling. Mahn's crew had saved the 21 structures at Few Acres, but the fire front roared on. Meanwhile, the fire was hitting the dozer lines around the Crandall base camp five miles away. Thirteen mobile homes and the store at nearby Painter Estates were going up. Three homes had been lost on Squaw Creek. The fire was encircling Crandall—it looked like they would have to pull out. Up the valley, in the other direction, fire was running in Onemile Creek and threatening to move down on another privately owned property, the Griscomb Ranch.

Six miles north of Few Acres, John Griscomb raced a pickup truck out of Cooke City and over Cooke Pass, heading for home. "I was furious," he remembers. "The Storm Creek burnout effort had been a complete disaster. Storm Creek itself had never left the park, and now the burnout was out of control."

The truck hurtled out of Cooke Pass and the Clarks Fork Valley opened in front of him. Flames and ribbons of smoke stretched for miles along the eastern slopes of the Absarokas. "Only yesterday," Griscomb says, "they had been the same grand forests I'd known all my life." He turned off the highway and crossed the wood bridge over the river. " 'This was it,' I thought. It was all over. The Storm Creek burnout fire was right behind me. The Clover-Mist fire was right in front of me. We were done for, cooked."

The ranch was nearly deserted. His family had left, driven off by the choking smoke. All the guests were gone. The stock had been moved downcountry. The ranch boss and his crew had evacuated their wives and children. "I saw the flames in the creek drainage above the ranch," Griscomb says. "The fire was a mile, maybe a mile-and-a-half away."

A team of engines from California was deployed near the main lodge. Griscomb spoke with the ranch boss. The firefighters, the ranch boss said, were talking about a backfire operation in the valley, but the plan was unclear. No one seemed sure what was going on. Rumors were flying. There had been a report that the Old Faithful Inn had gone up, killing the incident commander and observers on the roof. Someone said a supply helicopter had crashed. They watched the trees flaring up on the mountainside. At 5 P.M., the wind died down.

"We knew the fires would be on us soon," Griscomb says. "That night, maybe the next day." They spent the evening loading the trucks and preparing for the final ride out. The dismal task over, they stood outside and watched the fire in the hills. "No one spoke of the future," says Griscomb.

The next morning, Bob Mahn and David Anderson appeared with the Yellowstone crew, bringing news. They had saved Few Acres. They would try to do the same for the Griscomb place. The Old Faithful Inn had been spared. If they could hold out a few more days, it seemed likely that fresh resources would arrive. With the fires now dominating the national news, President Reagan had dispatched a Cabinet-level inspection team to Yellowstone, a move certain to increase the number of troops on the fire line. In Congress, the House and Senate had rushed through a bill, long opposed by the State Department, allowing Canadian firefighters quick entry to the country.

That was the good news. The bad news was that the fires weren't going to wait for the military and Canadian reinforcements. Another dry cold front was pushing into the region. It would arrive in about thirty-six hours, sometime around the dawn of September 10, maybe sooner.

Area Command had just issued another confidential Fire Behavior Evaluation. "The noteworthy issue now," the report stated, "is that the dramatic fire behavior on September 7— flame lengths 100–200 feet above the tree canopy, fire whirls, crowning and torching fire runs, fire-induced tornado-type winds—is only a prelude to conditions that are getting in place to produce a whole new magnitude of fire behavior on Saturday, September 10. We know the fuels are cocked and primed. Added to the explosive fuel situation on Saturday will be southwest winds 20–40 miles per hour with gusts to 60 miles per hour at the ridges." The coming firestorms, the evaluation warned, "will be massive in scale with tornado-like winds, fire whirls, and the potential for long-distance spotting up to 6 miles."

The winds, some fire analysts feared, would create a massive single fire complex. "The merging of independent fires or fire fronts," the report cautioned, "will greatly accelerate burning rates above that characteristic of an individual fire." Translated into the language of the fire line: the son-of-a-bitch was gonna rip through *everything*. "Direct structure protection" under the conditions would be "totally untenable." Firefighters should simply stay out of the way. "The only good safety zone will be a large black one."

Anderson had some additional bad news for Griscomb. The firefighters had learned that Clover-Mist Incident Command was going ahead with the backfire operation in the canyon. They were going to set fire to a swath of forest some ten miles long, from Pilot Creek to Squaw Creek, along the flank of the mountain. The supervising officer and the California strike team leader had the orders in their day-shift plan and insisted on following the command. Anderson and Mahn countered that the latest weather activity reinforced the argument that unpredictable winds could blow a backfire out of control. "Hell," Dave Anderson recalls saying, "just look at Cooke City!"

Anderson refused to participate in the backfire, and the dispute continued until a ranking member of the overhead team was called in. Mahn related the concerns to the overhead officer. The two local men had fought fire in the region for over twenty years and had firsthand experience of working in the valley, Mahn said. A backfire under the shifting wind conditions could blow fire all over the map.

The Yellowstone crew had another option: if they stood their ground at the dozer line above the Griscomb Ranch when the fire came down the mountain, they might be able to hold the line. The prevailing southwest winds would continue to blow fire across the mountain side and there was a chance they could prevent the gusting cross-winds from pushing the fire over the line. The California strike team could stand by at the ranch for structure protection.

The overhead officer changed the order, putting Mahn in charge.

The dispute settled, there were no hard feelings among the firefighters. They had a job to do and they got busy doing it. By noon on September 9, the California strike team and the Yellowstone crew had organized the structure protection and reconnoitered the defensive positions.

It was a tough situation. The ranch was nestled under the mountain slope, with the Pilot Creek drainage to the north and Onemile Creek to the south. Its inner compound—main lodge, barns, guest cabins—was scattered with old spruce and lodgepole pine. The Yellowstone crew would deploy along a dozer line cut through the timber 200 yards above the ranch.

The California team would stand by for structure protection inside the compound, pulling back to the hay meadows below the ranch if they lost the buildings. John Griscomb and the ranch hands would stay on as long as possible, but agreed to evacuate when their presence might hamper the firefight.

All afternoon the wind held off and the firefighters waited. At dusk, Griscomb loaded the last horses into a trailer and secured his belongings in the back of the pickup. "The sun set," Griscomb recalls, "for what seemed to be the last time over all that my family had built in the valley. We gathered in the kitchen of the main lodge, discussing the evacuation plan and trying to piece together the fragments of information that had reached us."

At 9 P.M., the phone line went dead and a moment later the power went out, plunging the kitchen into darkness. Outside the window, fire glowed on the hills. Then the flames began to climb. The big wind was coming back.

"That was it," Griscomb remembers. "We were outa there."

Mahn's crew had deployed along the dozer line and the California team had taken up their positions in the compound. It wouldn't be long now. The forest began to sway in a warm breeze.

By 10 P.M., Griscomb was driving over the bridge and out to the highway. They would head over the Beartooth Pass and down to Red Lodge. He drove slowly into the night. Above the ranch and all down the flank of the Absarokas the fire was picking up in bright pockets and wind-swept lines of flame, a beautiful sight, Griscomb recalls, yet vaguely evil. As the truck climbed the road, Griscomb saw a sudden light, a single tree exploding into fire on a distant black mountain—a spot fire, swept in on the wind. The ranch hand sitting next to him broke the seal on a fresh bottle of Jack Daniels and passed it over. Griscomb pulled hard at the whiskey. Soon, the burning valley was behind him.

*   *   *

196   Meanwhile, the Yellowstone crew waited on the dozer line. Mahn was in radio communication with the California team in the meadow by the ranch below. At midnight, the waiting ended.

"We're getting some wind here," Anderson remembers Mahn radioing the California team leader. Ash flew out of the gloomy woods and trees bent in the increasing wind.

The forest rumbled. "It's coming at us," Mahn said. A gust blew down the mountain, then a cross-slope wind shear hit them. They were going to get it from both directions at once. "Gusting winds," Mahn reported. "Wind shears."

Above the dozer line, wind-fed spot fires climbed into the canopy. A garish crimson glow lit the woods. Downslope winds were pushing the fire closer. The swirling red smoke came down on them. Embers flew from the night.

"All of a sudden" relates Anderson, "there was a large wind shift. The trees whipped like fly rods. Showers of sparks blew across the line. The fire was coming right at us."

From the meadow the strike team leader saw the fire explode and crown, then rush downhill. Anderson remembers the scene well.

"Get out! Get out! The fire is coming over you! Get out of there!" the strike team leader shouted in the radio.

Some of the jittery Yellowstone crew looked behind them, eyeing the long escape route through the thick forest.

"Hold your ground," Mahn boomed into his radio. "Watch for spots," he said, steadying the crew.

A spot fire torched the brush and began climbing into the trees across the line. Two women on the crew threw shovelfuls of dirt and knocked it down, while others with backpack pumps killed the fire. Then more spots flared. Then the wind shifted back, parallel to the line. A reprieve!

The night dragged on. The winds continued to shift. At 4 A.M., the fire blew over the line and spread into the meadow by the ranch. The California team caught the blaze and stopped it.

By now it was nearly dawn—Saturday, September 10. David Anderson shivered in the chilly air. When the fire didn't get them, the ice would. The humidity had begun to rise, dampening the flames. "It seemed likely," Anderson says, "that more of our squad would fall ill in the following days, and all but certain that the cold front would bring nothing but more wind and misery." Anderson peered through the smoking woods, making out the dim shape of the lodge below. "We had won another battle, but we still were losing the war," he says. "How much longer could we go on?"

XIX

Mammoth Hot Springs

As the Yellowstone crew fought off the Clover-Mist fire in the pre-dawn hours of September 10, Dan Sholly was preparing for the final defense of Mammoth Hot Springs. Seventy-two hours earlier, Sholly had burned out a patch of forest and meadow 8 miles south of park headquarters in a last-ditch attempt to divert the fire. The maneuver failed. Since September 7, the unchecked North Fork fire had grown to over 260,000 acres and spread 17 miles north and east, tightening its grasp around the capital of the Yellowstone Nation.

The tremendous irony of the situation was not lost on the chief ranger. All summer, dutifully suppressing his own growing doubts, he had defended the natural fire policy and the firefighting effort. Of course, that was the unspoken part of the Park Service mission: protect the organization and its policies at all costs. He had done that. The Marine veteran was nothing if not a good soldier. But had he fulfilled his duty to protect and preserve "the resource"? Misgivings had assailed him as he watched fire perimeters grow. "Now,"

200   Sholly recalls, "the fire had reached the homeland. It was right at our doorstep."

The top cop of paradise himself was under heavy assault. It seemed, Sholly wrote in his memoir, "as if the world were falling in around the park." Community meetings had become vicious, jarring affairs. The locals treated him like a "stormtrooper." The media barraged him with "idiotic questions." Political pressure was intense. Even as he prepared for the last stand at Mammoth, his attention had been diverted by the arrival of the presidential "fact-finding" delegation, led by Interior Secretary Hodel and Agriculture Secretary Richard Lyng. Within hours, Hodel would back away from his earlier strong defense of the park, now referring to the fires as "a disaster." The remark would make more headlines, keeping the fires the lead story in the nation, a position it had occupied since North Fork's run at Old Faithful. The reporters with their laptop computers and satellite-feed trucks had dashed up from Old Faithful and now were poised at Mammoth, hungrily awaiting the final debacle. Sholly viewed them with disgust bordering on hatred, a feeling shared by many in the park.

But Sholly had bigger problems than the questions from the media, and not just at Mammoth. The new finger of the North Fork fire, born with the fire's run at Old Faithful, was pushing across the Central Plateau toward the elegant, neoclassic Lake Hotel at the tip of Yellowstone Lake. It seemed the nightmare would never end: today's target of destruction was Mammoth; tomorrow's was Lake Hotel. A triple threat was closing on the Lake complex: North Fork was bearing in from the west; the Clover-Mist fire was backing down from the east; and the Snake River fire was threatening to move north along the rim of Yellowstone Lake.

Clover-Mist Incident Command was pressuring Sholly to let them put dozer lines across the Pelican Valley to hold off the fire. But Pelican Valley was premier grizzly bear habitat and Sholly didn't want the dozers there. A ranger at the East Entrance had delayed dozer entry to the park and a story about the fracas had already gone out over the wires. "Another controversy!" Sholly recalls with some bitterness.

"Stories went out saying we were 'preventing firefighters from entering the park.'"

"The fires had taken control of our lives," Bob Barbee remembers. "Not in our wildest dreams would we have imagined that North Fork would travel over 40 miles, or that the small isolated fires of the Snake River Complex would one day threaten Lake Hotel, spread across the Continental Divide, combine with the Huck and Mink fires, and stretch over 30 miles. Those sorts of fires were unimaginable in July."

What had they done wrong? Should they have been more alarmed by the extreme fire behavior at Calfee Creek on July 14? Had they missed the signals: the drought, the low-humidity numbers, the unusual fire movement through young lodgepole stands? "We all thought it would rain," says Dan Sholly, "It always had. But if I had known, for example, that the little Clover and Mist fires would merge and burn over 31 miles, of course I would have done things differently."

Too late! It looked like they were all going down now. Barbee and Sholly; Lake Hotel and Mammoth Hot Springs; probably Old Faithful too, still threatened by North Fork, which had sent a second firestorm past the north end of the complex on September 9; probably Canyon Village too; probably Cooke City and Silver Gate and West Yellowstone too. It looked like everything was going to burn up. The strong dry winds would never end. The firefighters on the lines were dazed and near defeat. Injuries and sickness were skyrocketing. It was too much for Sholly to think about. His career would be ruined.

Today was to be the day of the ruination of his career. The latest news from the fire gods at Area Command was that disaster was written in the wind. *Another* windstorm was on the way. "The dramatic fire behavior of September 7 and 9," said the newly issued Fire Behavior Evaluation No. 3, "is only a prelude to conditions that are getting in place to produce a whole new level of magnitude of fire behavior today. In fact, this system will equal or exceed the big blow of Saturday, August 20."

It was time to circle the wagons at Mammoth Hot

Springs. The cavalry was nowhere in sight. No one was going to ride to the rescue. They had lost. They had failed to stop the fires. Yellowstone Park, in effect, had been burned down by an idea. Here was a new kind of politics at work. The politicians of environmentalism—the scientists, bureaucrats, and special-interest groups—had carefully nurtured the ecosystem paradigm and brought natural regulation and natural fire into national policy. This, for good or ill, was their reward. For the guardians of Yellowstone National Park, the day of reckoning had arrived.

After a restless half-hour nap, Sholly woke at 3:30 A.M. on September 10 to the sound of fire. Bunsen Peak—a mile-and-a-half away from the 300 structures of Mammoth Hot Springs—was covered in a heaving blanket of red and gold, and flames were creeping along the ridge above Mammoth, toward Sepulcher Mountain. At 4 A.M. he walked over to the operations center and met with Steve Frye. They had held off evacuating their families from Mammoth—the hotel had been emptied a few days earlier—but could put off the decision no longer. It filled them with an obscure sense of disgrace. For the first time in its 116-year history, Mammoth Hot Springs, the old Fort Yellowstone, was being evacuated. With the evacuation, Yellowstone Park would for the first time be closed. Throughout the summer, Bob Barbee had managed to keep at least one or two of the roads and entrances open. It was a matter of honor as well as economics. Now the situation had finally grown too complex, too threatening, too dangerous. Sholly walked over to Barbee's home and gave the superintendent his recommendation.

The evacuation signal sounded at 5 A.M. Bob Barbee stood outside in the chilly dawn and watched the last civilian refugees of the Yellowstone Nation take the only safe road out, north for Gardiner and Bozeman. "It was an incredible sight," Barbee recalled. As the cars slowly wended down the plateau from Mammoth, a vast tempest of smoke and fire began to move in from the south. "There was a certain majesty to the whole thing," Barbee says. "You looked to the south sky and it seemed that an immense storm was on the way. At that awesome moment, the power of nature over man reaffirmed itself."

Sholly met with his top firefighters at the Mammoth command post. North Fork had turned into a voracious beast. East of Mammoth, North Fork's Wolf Lake portion was close to the Clover-Mist fire. On September 9, another finger of North Fork/Wolf Lake had run hard north, hitting a 10,000-acre burnout on the south flank of the newly merged Hellroaring and Storm Creek fires. Tenuous bridgeheads, thus, were linking—or on the verge of linking—North Fork to Clover-Mist, Hellroaring, and Storm Creek. Park headquarters was now cut off to the east and south by North Fork, and there was no road west. The only escape was north, through Gardiner and out the Paradise Valley.

That route was about to be cut off by a menacing finger of the North Fork fire—a "finger" that had burned 25 miles north from the road between Madison Junction and West Yellowstone during the past two weeks. The fire had taken the rough form of a giant hand, with five enormous fingers spreading 15 to 40 miles from the great black palm of the Madison Plateau—a thumb east toward Lake Hotel; a finger northeast to Canyon Village and curving up to Tower Junction; another running north by northeast from Norris Junction, below Arrow Canyon and into the Washburn Range; another moving east of the Norris–Mammoth road onto the Blacktail Plateau; and the fifth holding mostly west of the Norris road, climbing north for Mammoth while its rump end continued to hammer at the lines outside of West Yellowstone.

The birth of this giant haunted Dan Sholly and other officials at Mammoth. Could they have stopped it that first day on the Madison Plateau? Should they have allowed the dozers in on July 22? If they had stopped North Fork, the threat to Mammoth would be nothing but a bad dream.

That bad dream, the fifth finger of the North Fork fire, was about to flick park headquarters off the map. The forecasted windstorm would turn its leading edge into a much wider flame front. Thick smoke was gathering over the com-

204 pound. The fire was closing in. It was splitting into the narrow wooded drainages and passes surrounding them. It might come roaring out from anywhere when the winds picked up: straight off Bunsen Peak, or out from Golden Gate Pass, or out from Snow Pass, or down Glen Creek. Maybe it would back down from Sepulcher Mountain to the northwest. Maybe it would slither northeast, sneak along the back of Mount Everts and surge out from behind them, cutting off the escape route to Gardiner. Maybe it would do all those things at once.

Sholly and Frye decided to deploy some of their 300 troops to the perimeter, continue fireproofing, hosing and foaming the buildings, and stand ready to fight the fire wherever it blew in. "Frye brought me some more bad news," Sholly recalls. "Area command was denying us air support. There was too much smoke. I had been counting on the air support to hold the fire at the perimeters."

The two men had over forty years of firefighting experience between them and they knew what lay ahead: late morning would give way to afternoon, the temperature would rise, the humidity would plummet, the winds would increase, and the place would blow up.

Around noon, the wind strengthened. Waves of smoke rolled off the hills, blinding the firefighters, and the sound of exploding trees came near. The fire had snaked around and was moving in from the west.

A radio report came from the perimeter: "The fire was coming out of Glen Creek Canyon, at the Youth Conservation Corp building complex," Sholly says. He raced to the area. More reports flooded in over the car radio. The fire was pushing past Golden Gate! The fire was moving off Bunsen and along the Gardiner River! The fire was blowing down from Sepulcher Mountain!

At the Youth Conservation Corp site, "downdrafts from side canyons were throwing out long tongues of flame," Sholly says. Crown fires burst in the trees 50 yards from the buildings. "The heat, the smoke, was overpowering." A phalanx of engines poured water on the blaze. The firefighters stubbornly held their ground, their faces bent away from the inferno.

A blast of wind threw the fire at them. Behind the chief ranger, spot fires were leaping up in the meadow by the horse corral above the Mammoth compound. It was an awful moment for Sholly. He would be disgraced, dishonored. It was Saigon all over again—his own personal Vietnam!

Sholly turned the car around. He had to get back to the main complex. He had to be there for the end. And at that moment, perhaps minutes away from the destruction of Mammoth Hot Springs, maybe only hours away from the consolidation of all the Yellowstone fires into a single gigantic blaze, the Great Spirit smiled on them at last.

"I couldn't believe it," says Dan Sholly. "I just couldn't believe it."

It started to rain. The new cold front was carrying moisture—not a lot, but enough. A misty drizzle began to fall over Mammoth Hot Springs, and over Old Faithful and West Yellowstone, cooling the North Fork fire. A light rain was falling over Mahn and Anderson in the Clarks Fork Valley, and over Bates in Canyon Village. Snow—cold wet beautiful snow—was drifting down on Silver Gate and Cooke City, drifting over spike camps high in the Absaroka Range, over Storm Creek, over Hellroaring. Snow was mingling with rain at Snake River, at Huck, and at Mink.

Abruptly, it was over. They had been plucked from the brink of a cataclysmic defeat. There would be more battles ahead—the dangerous business of mopping up, fights over policy, assigning blame and assessing damage—but suddenly, what nature had started, nature was stopping. Winter was coming to the high country.

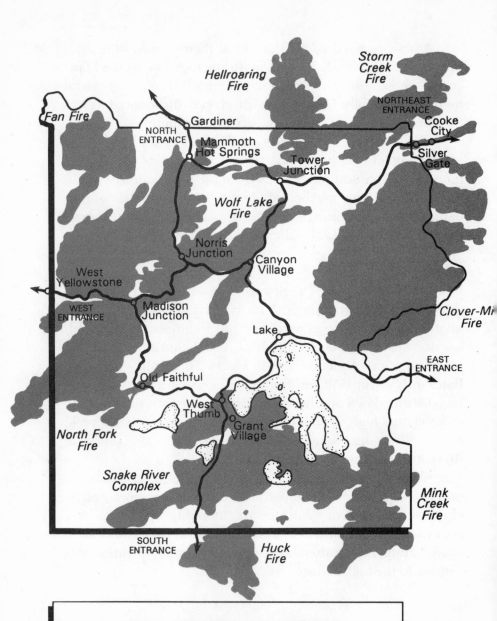

YELLOWSTONE FIRES
September 11

Fire perimeters cover 1.2 million acres.

© Linda Marston, 1993

Epilogue

Mammoth Hot Springs
March 1993

The mixed snow and rain of September 10 and 11—precipitation ranging from less than one-tenth of an inch at Mammoth to three-quarters of an inch at higher elevations—was enough to give the firefighters the upper hand, enabling them to begin to establish secure control lines and make the transition to mop-up work. The cold wet dirty business of mopping up would continue into November, when heavy snows extinguished the last embers of the great blazes.

For the country, it had been the worst fire season since 1924, when an estimated 30 million acres had burned; in 1988, a total of roughly 6 million acres had been consumed, much of it in Alaska. Final surveys of the Greater Yellowstone Area indicate fire zones of about 1.2 million acres in the hopscotch "mosaic burn" pattern. According to Yellowstone Park's final burned-area survey, issued in December 1989, the "total acreage affected by fire" inside the park was just under 800,000. The final figures from Yellowstone marked a significant downward revision, chiefly because unburned areas within the "mosaic" had been factored out of the burn counts. The survey reported that 323,000 acres of the affected area were "canopy burns," in which all trees were killed, and 280,000 acres were "mixed burn," a combination of canopy burns and surface burns in which some trees survived. Sage and grassland burns, isolated stands of trees, surface burns under unburned canopies, and a variety of other burn types made up the rest of the affected acreage.

Over the course of the fire season, twenty-five thousand firefighters passed through the Yellowstone country. Total cost of the fires was over $120 million. At the peak of the battle, during the first ten days of September, 9,500 firefighters and 117 aircraft were working amidst powerful winds, low visibility, major fire runs, and falling trees. "Safety first," the guiding principle of wildland firefighting, handsomely paid off at Yellowstone in 1988. That no one was killed during almost three months of intense work is a remarkable testament to contemporary safety practices.

Unfortunately, the mop-up stage would bring several fatalities. Pilot Don Kuykendall was killed in a plane crash on September 11 in Jackson, Wyoming, while transporting fire personnel. On October 11, Clover-Mist firefighter Ed Hutton died when a snag came down on him. And in Cooke City, a hermit named Tommy Garrison was burned out of his cabin by the Storm Creek backfire. Bewildered and broken by the loss of his home, he died soon after.

Lower down on the food chain, wildlife fatalities and land damage were not as great as many had feared. Someone in Cooke City took pity on a black bear discovered with its paw pads burned off and dispatched the animal, but most of the large animals simply moved away when fire neared. Yellowstone Park reported locating 243 elk, 4 deer, 2 moose, and 5 bison carcasses within fire perimeters. Most of the elk fatalities were due to a single incident on the Blacktail Plateau, where part of the herd became trapped by a flank of the North Fork fire and died from smoke inhalation. By contrast, the harsh winter of 1988–89 killed about 5,000 of Yellowstone's 20,000-head elk herd—natural regulation at work, culling the weak, the sick, and the old.

After some initial problems with erosion and siltation,

Yellowstone's blue-ribbon trout streams seem to have bounced back. Indeed, by 1992 there were some reports indicating that the trout had grown slightly larger, perhaps because of nutrient transport of forest ash into the aquatic systems. Although Yellowstone's affected forests will not see mature trees for at least sixty to a hundred years, studies conducted in the four years following the fires indicate that lodgepole pine and aspen saplings, as well as a wide variety of undergrowth, are coming back. As for the possible "soil sterilization" reported in the aftermath of the fires, Park Service and Forest Service officials say that only tiny parts of the Yellowstone area are seeing no regrowth at all. Of course much can happen in the coming years, including reburns that would destroy both saplings and the regenerative capacities of the area, but early surveys tend to confirm park claims that, in the words of Chief of Research John Varley, there was "no ecological downside" to the fires.

Yet despite the passage of time, the Yellowstone fires continue to raise questions and tempers, particularly in the Northern Rockies and in the community of wildland firefighters. Something went wrong, people believe. Yellowstone National Park's spectacular beauty has a darker, more ominous edge to it now, and those concerned with Yellowstone's fate—indeed with the fate of all of America's wildland preserves—continue to seek answers from the spectral forests. Two questions seem to overshadow all others. Could the fires have been stopped? And *should* the fires have been stopped— that is, was the "let-burn" policy, the natural fire component of the natural regulation policy, a good thing?

Could the fires have been stopped? In response to the national uproar, calls for resignations, and demands for congressional investigations, the Park Service and Forest Service embarked on a series of internal studies. One key document was a two-hundred-page report issued by the Greater Yellowstone Coordinating Committee, the *Greater Yellowstone Area Fire Situation 1988*. Known as "the Neckels Report" for National Park Service Deputy Director Jack Neckels, chairman of the fire

210   task force, it was a quickly put together, exhaustively detailed—if not always 100 percent accurate—chronology of all the fires. The second key document was a series of six internal "fire management reviews" conducted by Park Service and Forest Service officials, along with other agencies involved in the fires. Five of the reviews examined the five major fire complexes: Snake River, Huck-Mink, Clover-Mist, Storm Creek–Hellroaring, and North Fork–Wolf Lake; the sixth review examined the role of Area Command.

Although some senior Forest Service personnel have privately criticized the fire reviews as a "whitewash"—the fox reporting on the chicken coop, as it were—the Neckels report combined with the management reviews provides a reliable, if somewhat patchy, account of successes and failures. This writer used the two studies as a starting point for four years of research into the fires, including over a hundred hours of on-the-record interviews and the close examination of over 3,000 pages of documentation, most of it from the official fire archives at Mammoth Hot Springs, and a small portion of it provided by other sources, including sources requesting anonymity.

Could the fires have been stopped? Perhaps the greatest single problem in fighting the Yellowstone fires was the multiplication of command levels. Area Command never fully established its authority and thus, in the words of the Area Command management review, could not "effectively coordinate strategies and critical resources among incident-management teams." This was not the fault of the Area Commanders themselves but a failure of the command and control apparatus, the Incident Command System. The system—designed to be used in other emergency situations as well, such as earthquakes—could not handle its 10-million-acre area of responsibility, a region with many intersecting jurisdictions and strong-willed personalities. Area Command did not have the political weight to go up against two park superintendents, six forest supervisors, and numerous regional managers from the Park Service and Forest Service. "Area Command was never given clear directions from above," says one Area Commander. Power struggles consumed far too much of Area Command's time. This is an ill omen for the future, particularly for areas like California, where earthquakes and forest fires will occur within complex webs of jurisdictional authority.

One answer might be the creation of a post of a single "fire czar," to be granted virtual—albeit temporary—martial law powers in emergency situations. The law also could mandate the immediate assembly of an advisory panel of all the major regional players, as well as ordinary citizens, thus assuring that the fire czar would have the input of those most familiar with the area. It is unlikely, however, that a powerful centralized command structure alone could have stopped the Yellowstone fires. Each fire was unique. But a pattern of problems emerges when all the fires are considered together.

The Snake River Complex, for example, eventually covered 224,000 acres and cost the taxpayer over $10 million to fight, according to the Neckels report. But "Snake River" actually was a combination of the Falls, Red, and Shoshone fires, with some additional later starts. The early history of these fires suggests that they were not moving fast and that with the appropriate resources—and an immediate-suppression policy similar to that followed by the Forest Service in nonwilderness areas—they could have been stopped within the first weeks of ignition.

The Snake River fire review obliquely concedes this, noting that "the concern of the incident commander in not 'putting the fire out' " was echoed by other incident commanders in 1988. Of course, well into July park policy was *not* to put the fires out. This "limited-objective strategy" of letting fires burn and protecting lives and structures, the Snake River review records with wild understatement, "is not universally accepted within the fire-management community."

It's doubtful that the fires south of Yellowstone Park and the Snake River Complex—Huck, Mink, Hunter, and Emerald—could have been stopped by an aggressive early attack. Given the high winds on the days these fires started, the only one there is any question about is Mink. The Huck-Mink fire review noted that Bridger-Teton National Forest Supervisor

212 Brian Stout made a "high-risk" call when he allowed the Mink fire to burn for three days as a natural fire, before declaring it a wildfire and taking limited, "light-hand" measures to contain it. By the time the snows fell, the Huck-Mink complex covered 225,000 acres and cost an estimated $14 million.

Behind the scenes, Forest Supervisor Stout was the subject of much sharp criticism for refusing to "integrate" with Area Command and going his own way during the summer of 1988. "Stout was never a team player," notes a Park Service official. Stout bypassed Area Command, routing his resource requests through the Forest Service's Region Four offices in Ogden, Utah. He did not appear at many Area Command meetings—meetings criticized by some incident commanders as a waste of time. Both the forest supervisor and his apparently handpicked incident commander, Dale Jarrell, "experienced many problems" with Area Command, the Huck-Mink fire review noted. In another mark of the multiplying command levels, Stout and Jarrell established their own "mini Area Command" on September 1.

Also distant from the Area Command control efforts were the Storm Creek and Hellroaring fires. By the time the snows fell over Cooke City, Storm Creek fire perimeters stretched over 107,000 acres and firefighting costs were up over $8 million. Hellroaring covered 81,000 acres and cost $4 million before winter stopped it. Hellroaring, born in high winds from a fire started in an outfitter's camp on the eve of Black Saturday, could not have been stopped. Yet, with the luxury of hindsight, it seems certain that the Storm Creek fire, burning virtually unattended from June 14 to August 23 (except for monitoring efforts and a six-day Type I assault beginning on July 4) could have been stopped with a sustained aggressive attack up until shortly before Black Saturday. But Storm Creek was a low-priority fire, and it was burning in a wilderness area that allowed for natural fire.

Storm Creek was Yellowstone's forgotten fire, a blaze rich in ironies. It was a "positive" natural fire that caused much human distress. Its mysterious southward lunge on Black Saturday confounded the fire gods. It gave birth to the controversial burnout plan that nearly destroyed Silver Gate and Cooke City. Yet the burnout—which may have indirectly caused the death of hermit Tommy Garrison, devastated the "visual resources" north of the hamlets, and destroyed ten cabins and seven storage structures east of Cooke City—was a good fire-fighting plan. It *did* appear that the Storm Creek fire was poised to drop out of Pebble Creek and into Soda Butte Creek, thus creating a serious threat to the two communities; according to Interior Department officials, aerial surveys show that a tongue of the Storm Creek fire did in fact move close to the gateway communities, linking into the north end of the burnout zone.

As with all the fires by early September, dwindling resources, communication problems, and confusing command structures dogged the Storm Creek and Hellroaring efforts. "There was a lack of communication between the ICs on the Hellroaring and Storm Creek fires and Area Command in West Yellowstone," the Storm Creek fire review said. The creation of a "unified Command Team" in Cooke City—to deal with a situation in which the Storm Creek command was in charge of the wildfire but the Clover-Mist command was in charge of structure protection—caused "confusion," the fire review mildly noted.

The Hellroaring portion of the fire review also pointed to another source of confusion, one that can be found in all the fires: the problem of terminology. When a fire was fought in Yellowstone Park or in a Forest Service wilderness area, it always was fought under "light on the land," or "minimum-impact suppression" dictates. But the firefighters themselves—traditionally in the business of *suppressing* blazes—often were unsure exactly what the terms meant. Usually it meant no use of bulldozers or other heavy mechanical equipment. But sometimes it was translated into the tactic of confinement—letting a fire burn to a natural geographic barrier such as a lake. (And, as we've seen, "confinement" itself could be interpreted as a way of implementing a natural fire policy.)

And sometimes "light on the land" meant no use of chain saws, or no cutting of low limbs, a tactic to deny surface fires

214   the "ladder fuels" needed to climb into canopies. And sometimes it meant no camps in the backcountry, or no food or
garbage in camp, because of bear danger. And sometimes it
meant no driving trucks across meadows. "There needs to be
better understanding," the Hellroaring fire review noted,
"as to what the definition 'light on the land' actually means.
Some interpreted it to mean 'no aggressive firefighting tech-
niques.' " The confusion in the ranks quickly spread to the
press and the public, which already was having trouble sort-
ing out "prescribed natural fire," "let-burn," "confinement,"
"containment," and other terms.

All these problems—multiple levels of command; coordination and communication difficulties between the various
command structures, the press, and the public; confusion
over terminology; and dwindling resources in the face of
increasing fire severity—are evident in the histories of Yellowstone's two biggest fires, Clover-Mist and North Fork. In the
case of Clover-Mist, it is entirely possible that communication
problems were a major factor in not establishing early control
over—and possibly even suppression of—a fire that eventually
covered 387,000 acres and cost $19 million.

Remember the events of the critical week of July 14. The
Shoshone National Forest's fire management officer agreed to
"accept" the Clover and Mist fires on July 14. Yellowstone
Park let the backcountry fires burn, turning its attention and
resources to Grant Village, the Snake River fires in the south,
and the Fan fire in the north. On July 21, Forest Supervisor
Stephen Mealey, returning from his fishing trip in the forest
with Vice-President Bush, reversed the decision of his fire
management officer and *refused* to accept Clover and Mist, by
then merged and burning at over 21,000 acres.

Many questions arise. Was Supervisor Mealey, distracted
by the VIP visit, not properly minding the store? Would a
prompt refusal on July 14 or 15 have triggered an aggressive
suppression response from Mammoth? Or would the Yellowstone Fire Committee—early in the game and still clinging to
natural fire—have dragged its feet on suppression anyway?
Furthermore, should the actions undertaken by Dan Sholly to

protect the Calfee Ranger Cabin alone have been enough to
initiate a wildfire declaration and suppression activities, as
some fire-management officials believe?

The answers are lost in the gray areas of fire-management policy. But the Clover-Mist fire review did conclude that "the park could have suppressed both fires with initial action forces" up to July 13. After that, "the number of strategic options [decreased] to confine and partial containment on July 23. . . . Options after July 23 were limited due to extreme fire behavior, low fire priority, lack of suppression resources, and logistical support." Yet there seems to have been other windows of opportunity after July 23. Curt Bates' Type I team did not even arrive until July 24 and left on August 2, bitter over not being allowed to fight the majority portions of Clover-Mist inside the park. Bates later told the *Denver Post* that he believed he could have held Clover-Mist to 47,000 acres in the Yellowstone backcountry. Clearly, opportunities to stop the Clover-Mist fire were missed.

But could the Forest Service, with all its resources and a swift attack, have stopped Yellowstone's greatest fire? By the time it was over, North Fork, including its Wolf Lake branch, burned across more than 500,000 acres and cost over $36 million, according to the Neckels report. Controversy still rages about Yellowstone Park's decision not to allow bulldozers over its boundary to attempt to circle the blaze in the hours following ignition on July 22. High winds had quickly pushed the blaze off the Targhee National Forest and onto Yellowstone's Madison Plateau. Forest Service officials assigned to the early stage of the fire believe North Fork could have been stopped with a bulldozer line. "We could have put it out at 1,500 acres," North Fork's first Type I incident commander, Larry Caplinger, told Jim Carrier of the *Denver Post*. But, says Caplinger, "I received official orders to back off and let it go."

The North Fork fire review, however, concluded that "control of the fire was not possible because of the very dry fuel conditions and strong winds." A whitewash? It doesn't seem likely. Eyewitness accounts, including some witnesses not interviewed by fire review panels, confirm that North Fork

216   was moving fast and spotting heavily. With spot fires recorded up to a half-mile in front of the main fire, it seems highly unlikely that a bulldozer line could have successfully contained the blaze. The fire could not, under any reasonable scenario, have been stopped.

As the fire season moved into August and September, North Fork would prove the fire gods wrong time after time, wreaking havoc in nearly 25 percent of Yellowstone Park. "The fire behavior book was rewritten" after North Fork, noted the official review. "Fire behavior in 1988 was at levels easily compared to the Big Blowup of 1910."

Could the Yellowstone fires have been stopped?   First, it must be noted that stopping the fires *was not the policy* until it was too late. (North Fork, of course, was a human-caused fire and under the fire-management plan it should have been suppressed. But, like Hellroaring, it was deemed unstoppable from a firefighting, not a policy, perspective.) But if Yellowstone Park had not had a natural fire policy and instead had followed general Forest Service practices of prompt aggressive initial attack, it seems certain that the Falls, Red, Shoshone, and other early fires could have been stopped. (Some, such as the Lava fire, were in fact stopped.) The Storm Creek fire, ignited in a wilderness area of the Custer National Forest, also could have been stopped with early and sustained action. Early attack also would have suppressed the Clover and Mist fires, according to the fire review, but the outlook for suppression following July 14 is by no means clear. Mink, Huck, Hellroaring, Hunter, and Emerald could not have been stopped. As for North Fork, the best evidence suggests that suppression of the fire never was a realistic option.

Which brings us to the second great question of the Yellowstone fires: *Should* they have been stopped? Is natural fire—and by extension the notion of the natural regulation of the ecosystem—a sound idea? Or should man have been wrestling with fire from the very first ignition—Storm Creek, June 14—possibly preventing some of the big fires?

By late July 1988, Yellowstone Park and environmental

groups had largely abandoned the defense of natural fire in the face of mounting public hostility. The moment the rain and snow began to fall, however, environmentalists rallied to the support of Yellowstone and the beleaguered National Park Service director, William Penn Mott. Writing in the September 11 *New York Times*, M. Rupert Cutler, president of Defenders of Wildlife, condemned the calls for Mott's resignation because of his support for the natural fire policy. "A firestorm of political opportunism and incendiary rhetoric is racing through the treetops of Congress," Cutler wrote. "It's scorching good science, laying waste to common sense and leaving nothing in its wake but smoke and hot air." Natural fire would rid the park of overaged pine and promote greater plant and animal diversity and abundance. Yellowstone, Cutler concluded, "will be a healthier and infinitely more interesting place because of this exceptional ecological event."

Spokespeople for other national environmental groups echoed Cutler's view. "There's been a lot of inaccurate information passed around," Tim Mahoney of the Sierra Club told the Associated Press. "These fires are going to get away from you no matter what the policy is." "We think the calls for Mott to resign are ridiculous," added National Audubon Society spokesman Robert SanGeorge. "Politics should not intrude on biology."

But the politics of environmentalism—a belief system built on the idea of a naturally regulated ecosystem and deep hostility to a role for humankind in the wilderness—had long ago intruded on biology. Local opinion sharply diverged from the rhetoric put forward by the national environmental groups, rhetoric that would soon be echoed by the major dailies and the networks. In contrast, the *Billings Gazette* wrote on September 11, "We believe that Mott, Park Superintendent Robert Barbee, and Interior Secretary Donald Hodel must resign, not because of their policy but because of their ineptness at adjusting policy to reality." Natural fire was not inherently bad, the paper noted, but clinging to it in a time of deep drought had proved "disastrous."

Ideas had clashed with reality and had been revealed as

218   flawed. To many in the region, this had been obvious all along. "By June," noted the Red Lodge, Montana's *Carbon County News*, "talk of a drought was already rampant. . . . Temperatures were inordinately high. Humidity was woefully low. Soil moisture was nonexistent." Creating a policy framework in which fires were allowed to burn, the *News* said, was the work of "cloud-bound idealists . . . well-meaning people" who had tried "to put a plastic bubble over this nation's remaining wilderness. But they were too late. The Rocky Mountain West has been settled for more than a century now, and no portion of it can ever again be the way it was."

This attitude threatened not merely natural fire, but the entire preservationist concept of a naturally regulated ecosystem in which humans would play little or no role. Local and national environmental groups were quick to strike back. The Greater Yellowstone Coalition, which had taken an uncharacteristically low profile as the fires threatened the homes of its members and friends—"chickenshit" was how an annoyed Bob Barbee termed it—reemerged in mid-September, sending a letter to its supporters to remind them that "Greater Yellowstone is a fire-adapted ecosystem. . . . Greater Yellowstone depends on periodic fires for healthy functioning." Executive Director Ed Lewis and Program Director Louisa Willcox assured members that the GYC intended "to play a central role in a careful review" of the fires and urged them to remember that "now more than ever, it is important that we look to the future of the Greater Yellowstone Ecosystem."

By December, the battle over natural fire had shifted away from Yellowstone and to the corridors of media and government power in New York and Washington. "Outsiders," the Old West's mythic enemy, were asserting themselves again. In an influential article for the *New York Times Magazine*, "The Case for Burning," published on December 11, the distinguished environmentalist Peter Matthiessen defended natural fire. Reporting from an area of "black limbless spires" near Canyon Village, Matthiessen wrote that instead "of gloom at the seeming 'loss,' the vast charred prospect, there came instead a heady sense of earth opening outward, of mountain light and imminent regeneration, which made me recall how oppressive I once had found the south part of the park, with its monotone enclosing stands of lodgepole pine. Far from being stunned by the destruction, I felt an exhilaration and relief, as if Yellowstone Park, for the first time in a century, had gotten a deep breath of fresh air."

The Matthiessen article concluded with lengthy quotes from Greater Yellowstone Coalition Executive Director Ed Lewis in defense of natural fire. "In general," Matthiessen gingerly noted, "the Greater Yellowstone Coalition is a strong supporter of park policies and decisions, which might seem to support criticism that such private groups and the Department of Interior form a conservation establishment that tends to encourage Park Service stagnation." Park critics such as Alston Chase have put it in much blunter terms, charging that environmental groups and government agencies have established a kind of "revolving door" through which an environmental elite move in and out of government, defending their programs and preventing any significant reform.

If the Yellowstone fires are any indication of the strength of established environmental groups and policy, Chase's charges—only hinted at by a mainstream environmentalist like Matthiessen—are correct. Four days after the appearance of Matthiessen's article, a group of government investigators, the Fire Management Review Team, submitted its draft report on the natural fire policy to the Secretary of the Interior and the Secretary of Agriculture. The team concluded that the natural fire programs in national parks and wilderness areas were basically "sound," but needed to be "refined, strengthened, and reaffirmed."

During the public comment period that followed the draft report, the case for natural fire had many supporters. All of them came to the startling conclusion that the Yellowstone fires confirmed their existing beliefs. The fires, Greater Yellowstone Coalition Program Director Willcox told a congressional hearing in January 1989, "confirmed that Greater Yellowstone is an ecosystem" because they followed "the course of drainages and other natural, topographical, and geographical features." Yet, besides roaring up a few drainages at the end of the season, the troubling aspect of the Yellowstone fires was precisely that they did *not* follow natural, topographical and geographical features; the tactic of confinement was a complete failure. Willcox went on to make the fantastic claim that "by the end of the summer, the boundaries of the fires' perimeters began to clearly delineate the area we think of as Greater Yellowstone," a claim that is patently false.

And not only was Yellowstone "fire-adapted," Willcox noted, but so were the gateway communities, which actually had benefitted from the fires. "Many businesses had banner years," she told the hearing, "particularly motel owners, restaurant and grocery stores which helped house and feed the firefighters." The biggest threat to the ecosystem was not fire but "development activities" such as "oil and gas exploration, hardrock mining, timber harvesting, and roadbuilding." The real threat, in short, was Forest Service policies of multiple use.

The Matthiessen article, the Fire Management Review Team's interim report, and the diligent public relations efforts of the National Park Service and the Greater Yellowstone Coalition set the tone for what would become an overwhelming chorus of articles and television reports in the spring and summer of 1989 announcing the miraculous "rebirth" of Yellowstone Park. "As Snow Melts, Yellowstone's Rebirth Begins," noted the headline in the *Washington Post* in May 1989. "Yellowstone Lives!" announced *U.S. News & World Report*. "From Yellowstone Ashes, New Life and Approach," said the *New York Times*. The gist of all the reports was remarkably similar: Yellowstone's death had been exaggerated; the wildlife and geysers were fine; grasses were coming back strong; and lodgepole pine seedlings were showing signs of growth.

Fire was, as *U.S. News* put it in a May 15 rebirth piece, "as elemental and value-neutral in shaping the park's ecosystem as the wind, the rain, and the cold." No reporter even questioned the notion that Yellowstone was an "ecosystem." The "true threat," wrote Michael Satchell of *U.S. News*, in a widely echoed view, was "development" within "the 12-million-acre Greater Yellowstone ecosystem." Fire was not the problem at all. "The realistic fear," Satchell darkly concluded, "is that the tapestry of America's most ecologically pristine national park will be unraveled, thread by thread, as its creeks are poisoned by mining wastes, endangered grizzlies are shot as they stray over the ecosystem's shrinking boundaries, forest clear-cuts destroy scenic viewscapes, and the development squeeze turns Yellowstone into a big-sky version of Central Park."

The *U.S. News* article demonstrates that the Yellowstone fires had been a confusing episode for the national media from beginning to end. Most reporters did not have the faintest idea what was going on inside Yellowstone, and accepted at face value what environmentalists and the Park Service told them. Most creeks, for example, run out of mine-free Yellowstone Park, not into it; grizzly bears are a "threatened," not "endangered" species, and hunting is strictly controlled; the park has no clear-cuts; and rather than shrinking, the definition of the Yellowstone ecosystem has been magically expanding for years. In 1985, Greater Yellowstone Coalition documents estimated the ecosystem at about 6 million acres; by 1987, it had become 10 million acres; by 1988, 12 to 14 million acres; and by 1992, the environmental group was claiming that the ecosystem "encompasses 18 million acres."

As Alston Chase was perhaps the first to suggest, the ecosystem paradigm has moved out of the realm of science and into popular culture and public policy without serious challenge. The media unblinkingly accept the fantastic elastic notion of ecosystems—be they Yellowstone Park, Planet Earth, or somewhere in between. So, increasingly, do politicians. In 1993, newly appointed Interior Secretary Bruce Babbitt delighted environmentalists when he told the House Committee on Natural Resources that he would take "an early ecosystem approach to species conservation." In Montana, the Alliance for the Wild Rockies and the Wild Rockies Action Fund have been promoting the Northern Rockies Ecosystem Protection Act, which envisions a naturally regulated ecosys-

222   tem, free of development, stretching across five states and more than fifty million acres, with Yellowstone Park and other already preserved areas at its core. Representative Peter Kostmayer, Democrat of Pennsylvania, introduced the act in the House of Representatives on September 9, 1992.

The "pro-environment," or "anti-development," slant of the media is, of course, further evidence of its liberal bias. The generally uncritical tone of the coverage of the Yellowstone fires, with the exception of the dramatic days from August 20 to September 10, when clearly something was going wrong, supports claims of bias. "Development," such as logging, mining, and hunting, the press was reminded by environmentalists, is "bad." Environmentalists were the good guys. Thus, natural fire, supported by the antidevelopment environmentalists, must be good.

In fact, it can be argued that due to a liberal bias, laziness, and guilt, the national media missed the big story, misreported the 1989 follow-up, and may even have blown an opportunity to help stop some major fires. Remember that the national press corps departed in late July, after some light rains and optimistic outlooks from Area Command and Mammoth. But in the first two weeks of August, line firefighters and officers began to speak of unsettling weather conditions. The report of the fire-behavior experts started to crumble. Local residents started compiling evidence of erratic fire movement. Incident commanders and overhead teams, such as Poncin's and Bates', were troubled by the fire-management tactics. And some "low-priority fires" were picking up speed, Park Service and Fire Service rangers knew. All these were ripe sources for the press. Yet the press had disappeared. It is entirely possible that some good national reporting might have exposed some of these factors and brought political pressure to bear for increased resources, perhaps altering the histories of the Clover-Mist and Storm Creek fires.

But the press missed the boat. Following Black Saturday, as we've seen, the media returned, and they were not in a good mood, believing that someone had pulled the wool over their eyes. This provided a spurt of good reporting and increased the pressure for action. Yet with the fires out, the media again began to heed the organized voices of the environmental movement and rend its garments in guilt over "misreporting" the great blazes. They had not seen the "context," had not understood natural fire, had exaggerated the destruction.

"If the first job of the media is to convey accurate information," *Washington Post*'s T. R. Reid wrote in a July 1989 act of contrition, "When the Press Yelled Fire!," "then we failed in our job." Yet if the media are both a mirror of society and a conduit of information, then the press did accurately reflect the confusion *on the ground in Yellowstone* over fire policy in July. It simply was absent during most of August—its greatest error. And by and large it did a good job of covering the fast-moving, chaotic events of early September. (As noted elsewhere, local media outlets did a superb job in covering the story, though they too backed off a little in early August.) Contrary to Park Service and environmentalist complaints, a review of articles and network footage shows that most reporters were careful to note that the whole park had not been "destroyed" and to try to explain the natural fire policy.

What the press misreported was not the fires themselves, but the follow-up. Finding themselves on the "wrong side" of an environmental issue in the immediate aftermath of the fires, they sought to make amends with the spate of "rebirth" articles. "The media came back with ash on their foreheads," Bob Barbee noted with some satisfaction about the follow-up. He had reason to be satisfied. By the summer of 1989 it was clear that the Yellowstone hierarchy would survive and that the issue had been turned to their advantage. Yellowstone Park had come through a monumental test of natural regulation and the public had been "educated" about natural fire.

The fires, it turned out, had been good for the Greater Yellowstone Coalition too. GYC President Thomas McNamee told the *American Spectator* in August 1989 that the group's membership and budget had doubled, largely because of publicity from the fires. (According to some accounts, the GYC has healed its spat with Yellowstone, and at a 1989 GYC meeting at Old Faithful Barbee was given a standing ovation. Others inside Yellowstone say the park leadership remains deeply embittered by the general desertion of environmental groups at the height of the fires.) McNamee says the group will continue to fight multiple use and push for greater preservation and a greater federal role. "We're sold on the idea of the West as independent," McNamee said, "but the fact is, they suck at the federal tit twelve months a year."

The Yellowstone fires also gave a jolt to a loose conservative coalition just getting underway in 1988, the "Wise Use" movement, representing the multiple-use side of the environmental debate. "The Yellowstone fires were a wake-up call to Western conservatives," said Ed Wright, an editor at *Our Land* magazine. "We see natural regulation, natural fire, and preservation as a way to close off use of the forests. We're ranchers, farmers, lumbermen, miners—advocates of careful, wise use of our natural resources. If radical environmentalists look at earth as a virgin and man as a rapist, conservative environmentalists look at earth as a garden and man as a gardener."

Wright and the Wise Use people have a long way to go. Environmentalist assumptions are deeply woven into the national fabric. Natural fire is part of the story of natural regulation, and natural regulation is not simply some arcane Park Service policy but part of a web of complex social, scientific, and theological notions embodied, as Alston Chase has noted, in the quasi-mystical notion of a self-contained ecosystem. And while the Yellowstone fires were a triumph of environmental politics, of natural regulation, it would be wrong to view this triumph as merely a contemporary phenomenon. The politics of nature stretch back through the 1960s, through Teddy Roosevelt, John Muir, and Emersonian transcendentalism to the culture war between pagan animist beliefs and Judeo-Christian teachings. In a stroke of animism the Druids would have appreciated, the GYC's Louisa Willcox held a "wake" for fire-damaged Clover-Mist timber slated to be cut down and sold, according to several sources in the Clarks Fork Valley.

"Man shall have dominion over the Earth," says Genesis. "The Earth does not belong to man," says Chief Seattle in a popular environmentalist text, "man belongs to the earth." Dominion in fire suppression or communion with nature in natural fire? We are still choosing sides—and not only in the world of fire.

Should the fires have been stopped? In the end, both the government and the media answered "no." Natural fire survived.

In March 1992, following a three-year suspension of the natural fire program while new policy guidelines were devised—a period in which, incidentally, all lightning ignitions in Yellowstone were successfully suppressed—Yellowstone Park revived the program under a new wildland fire-management plan. The plan significantly upgrades the coordination mechanisms between the park and surrounding forests, builds in a number of fuel-moisture and other "trigger points" for a wildfire declaration, and reenforces special protective "buffer zones" near the gateway communities. "Basically," says Fire Management Officer Phil Perkins, "the latitude about when to decide to fight a fire has been taken away from the park and built into the policy."

There remain a few voices crying in the wilderness. Alston Chase has written that it is not natural fire that needs investigation but natural regulation itself. As for Yellowstone, Chase and others have suggested a policy of "prescribed fire"—not prescribed *natural* fire, but fires deliberately set in the early spring and late fall to reduce heavy fuel loads. Of course, this is not natural regulation but involves a greater emphasis on a guiding human hand, a policy Chase generally favors. Yellowstone, in Chase's view, is not an intact ecosystem. Man must actively maintain the illusion of primitive America. Although there is room in Yellowstone Park's policy for human-caused prescribed fire, so far the park has backed away from its use.

Some have suggested that the Yellowstone fires belong more to the past than the future. Writing in the *New York Times* in the wake of the devastating fires in Oakland, California in 1991, fire historian Stephen Pyne noted that the real wildland fire problem is in what has been called the "wild-

226   land/urban interface"—the places where shifting demograph-
ics have brought more people and homes into the forests.
Many fire-management experts agree. In Yellowstone, the
focus would then shift from Yellowstone Park—at the heart of
the Greater Yellowstone Area—to its periphery, where Forest
Service domains meet private lands.

That would be fine with officials at Mammoth Hot
Springs. Almost five years after the great blazes, they are getting on with their lives. The fires have not hurt park visitation,
which in 1992 topped 3 million for the first time. No heads
rolled following the great blazes, but there have been the
usual promotions and retirements. Assistant Chief Ranger
Steve Frye, one of Dan Sholly's top deputies, has gone on to be
chief ranger of Glacier National Park. Incident Commander
Curt Bates has become the supervisor of the Custer National
Forest. Incident Commander Dave Poncin has gotten out of
the fire game but remains in a top staff position at the Nez
Perce National Forest in Idaho. Others have moved on or
moved up. At the Yellowstone fire cache, Phil Perkins, pro-
moted in 1989 to fire management officer, is testing the new
fuel-moisture gauges. The fires, he admits, "are still controversial in the fire community. They certainly were the biggest
event in my career."

Over at the Administration Building, Public Affairs Chief
Joan Anzelmo has moved on. Due largely to her cool performance in 1988, she has been appointed director of external
affairs for the Boise Inter-Agency Fire Center, which in 1993
was renamed the National Inter-Agency Fire Center. "It was a
piece of history," Anzelmo says of the fires. "I'll never forget it,
but I hope I never have to go through it again." Next door,
Superintendent Robert Barbee is preparing for the months
ahead. He has no regrets about the fires. "I take full responsibility for everything," he says. Upstairs, Chief Ranger Dan
Sholly echoes his boss. "We're the ultimate authority. We're
responsible," he says with a smile. Sholly seems to be a man
at peace. He has published his own account of his life at Yellowstone, including the fires, and used the opportunity to settle a few scores with park critics. Sholly and Barbee have been

unfairly blamed for much of the fires. The two dedicated public servants stuck to the mission—natural fire—as long as possible, and then were outflanked by Area Command and
undone by the wind. All that was behind them now, however.
"We're looking to the future." Sholly says.

The future is sure to bring more controversy, as natural
to the world's oldest national park as the seasons. The big
issues of the moment are the reintroduction of the gray wolf
and a gold mine slated to go into operation near Cooke City.
Environmentalists support the wolf and fear the gold mine
will "impact" watersheds, grizzly bear habitat, and "visual
resources." Yellowstone Park agrees with both positions but
will wait for the politically correct moment to openly support
the wolf and oppose the mine.

Natural fire will be phased in slowly and watched carefully. Politics will not allow a repeat of 1988 in the near future,
although nature may have something to say on the matter:
enormous fuel loads remain in the Yellowstone area. "Nature
always bats last," says Phil Perkins.

In the end, the policy of letting fire burn must have been
something of a pyrrhic victory for men like Dan Sholly and
Phil Perkins and Dave Poncin. For what had been lost in their
paradise was not the natural beauty of the park—Yellowstone
remains staggeringly beautiful—but an idea of the West itself
as a defining American ideal, that frontier notion of capable
men pitted against the elements. The mystique of the Old
West had suffered a deadly blow. The New West—the West of
bureaucrats, interest groups, and scientists—had carried the
day by carrying natural fire to its awesome end.

And yet perhaps all is not lost in paradise. With the gradual easing of the natural fire restrictions, capable men and
women again had stood against the high and wild—had redis-
covered the true wildness in the wilderness. They had per-
formed, in Hemingway's shrewd definition of courage, with
grace under pressure. Twenty-five thousand firefighters had
endured a million acres of fire over three months. Sholly at
Calfee Meadow. Poncin at Grant Village and in pursuit of
North Fork. Perkins on the Thunderer. Hogan at Big Park

228   meadow on Black Saturday. Bates on Clover-Mist and at Canyon Village. Sears at Silver Tip Ranch. Liebersbach at Cooke City. Mahn and Anderson in the Clarks Fork Valley. They had snatched away some not insignificant victories, victories of the spirit if not always of the flesh. They had fought and they had survived. The frontier lived in fire.

Acknowledgments

Joan Anzelmo, now director of External Affairs at the National Inter-Agency Fire Center in Boise, Idaho, was a model of professionalism during her tenure as chief of Public Affairs for Yellowstone National Park. Anzelmo balanced her dedication to the park and its values with an honest attempt to grant all my requests for interviews, documents, and access to fire areas. The only request she flatly turned down was my first one: for a helicopter.

In Washington, Pamela Segal proved to be a trusted friend, advisor, and typist. In Montana, Dorothy Houser provided valuable research. In New York, Midge Decter taught me, among other things, that ideas matter—and gave me the freedom to report on the fires for the *American Spectator* while serving as her deputy at the Committee for the Free World.

Wladyslaw Pleszczynski, Bob Tyrrell, Ron Burr, and the staff of the *American Spectator* have provided me with an intellectual home for over a decade and have provided the country with that rarest of things, a magazine that celebrates the individual voice of the writer. My thanks to the *American Spectator* family, and its generous supporters, particularly the Sarah Scaife Foundation. Thanks also to John O'Sullivan,

230 John Fox Sullivan, and Marty Peretz for giving me my first breaks, and to Norman Podhoretz and Neal Kozodoy.

At Yellowstone Park, Superintendent Robert Barbee decided to give me full cooperation—no easy decision, given the highly critical thrust of my reporting—and signaled that others should do the same. My thanks to him, and to Dan Sholly, Steve Frye, Terry Danforth, Phil Perkins, Dick Bahr, Don Despain, John Varley, George Robinson, Rick Gale, Sandi Robinson, Marsha Karle, Cheryl Matthews, Steve Tedder, Randy King, John Donaldson, Joe Fowler, Curt Wainwright, Sue Consolo, Tom Tankersley, Bill and Debbie Young, Amy Vanderbilt, Paul Schullery, Lee Whittlesey, and Dick Clark and the 1990 Clover-Mist horsepack trip.

Among both Park Service and Forest Service personnel, there were some who requested anonymity. My thanks to them, as well as to all the Forest Service employees and others who responded to my inquiries, including Dave Poncin, Curt Bates, David Liebersbach, Larry Sears, Pat Pierson, Forest Service researcher Dave Peterson, Jack Troyer, Betty Schmitt, and Keith Crummer.

Many residents of the high country around Yellowstone—particularly in Cooke City, Silver Gate, and Red Lodge—were generous with their time and hospitality. I'm grateful to Gus and Margaret Hart, Mike and Beth Evans, Dave and Kathy Hannahs, Wade King, Michelle Perrin, Hayes Kirby, Jeff Henry, Ed Francis, David (Paco) Anderson, Dale Christianson, John and Anne Graham, Dave Majors, Garry Brown, Dan Hogan, Joan Humiston and family, Ralph and Sue Glidden, Darrell and Patt Crabb, Jim and Dusty Peaco, Doug Vivian, Jim and Suzanne Kadous, Jean Albus, Angie Hazelswort, Doug Hart, Ellen Hart, Louise Chandler Hart, Skip and Debbie Bratton, Kent and Diane Young, Gary and Jane Ferguson, Margi Sheehy, Kay King, Dennis and Chris Derham, the Ellsberry clan, Rick and Jenny Hooven, and Alison McKinley.

Thanks also to Lisa Middents and Megan Desmoyers at the Hemingway Collection in Boston; cowboy poet Ray Pendergraft; cartographer Linda Marston; Bob McHugh, encountered one smoky night in Cooke City; leading "geyser gazers" Paul Strasser, Suzanne Strasser, and Heinrich Koenig; Don Leal and his colleagues at the Political Economy Research Center; my friends and colleagues at *Insight*, particularly Stephen Brookes, Kirk Oberfeld, Richard Starr and Brig Cabe; John Michel; Larry Ashmead, Scott Waxman, and all the folks at HarperCollins.

Finally, love and lasting gratitude to my own secret team of family, friends, unpaid researchers, teachers, lovers and disputants: Catherine Morrison, Seth Morrison, Sylvia Orloff; Peter, Helga, Max, Alex, and Jessy Orloff; Dr. Elizabeth Kaufman, Rabbi Joseph Kaufman, Ezra, Eli, and Esti Kaufman; Rob Hart and his remarkable family; Nick Kneibler; Dana Peterson and Dr. Marjet Schoen; Peter and Simmie Issenberg; Wynn Miller, David Segal, and Chris Rosen; Daniel Masler and Kristina Hagman; Edward Masler; Susan Brown; Illia Barger; Josh Gelman; Stephen Sandy; Valery Levine; Jeremy Sager and Paula Clements; Nina Nilson; Amy Lumet; April Parker and Cindy Cordes; Dr. Brian Grobois; Susan Mendelson; Shuli Sade; Jon Ewing; Peter Delano; Anne and Elise Aronov, Stephen Horenstein, Bob Cohen, Uri Rubinstein, Hillel Kraus and Athanasius Gadanidis; Diana Laurence; Fran Shalom; Jaime Kelter; Pam and Ed Segal; David Trout; Joseph S. Murphy and Kay Fay; Jim Moffett and the Jones Cafe crowd; Alan Cheuse, Nick Delbanco, Harry Mathews and George Garrett; Tom Mangold; Stuart R. Shapiro; Eric Delong, Rod Williams, Rick Wendland, Clive Runnels, and the rest of the Kents Hill delinquents. This book is for all of you.

Glossary

Area Command: An organization established to ensure coordination, planning, and logistical support. *See also:* **Incident Command System.**

backfire: Method of indirect attack to remove fuels within control lines; often a broad defensive strategy; employed when fire is some distance from lines. Firefighters often, confusingly, use **backfire** and **burnout** interchangeably. *See also:* **burnout.**

backing fire: A fire or part of a fire spreading against the wind.

blowdown: An area of downed fuels, usually associated with high fire danger.

blowup: Sudden increase, often violent, in fire intensity and rate of spread. *See also:* **firestorm.**

burnout: Method of direct attack when fire is at or close to control lines; setting fires inside control lines to strengthen lines. Often used interchangeably with **backfire.**

confine, confinement: Technical term that caused much confusion among public. To put a fire "in confinement" is to allow it to burn within identified geographic boundaries. *See also:* **contain; control.**

contain, containment: Technical term that caused much confusion among public. A step up the suppression ladder from **confinement,** containment involves building fire lines, or control lines, around the fire and letting the fire burn itself out. Under a containment strategy, little or no work is done on the interior of the fire.

control: A step up the suppression ladder from **containment.** Direct attack; cooling off hot spots within interior of fire; going in and putting the fire out.

control line: Also called a fire line; interruption of fuels near fire, usually made by scraping away fuels down to mineral soil. Variations: **handline,** made with hand tools; **wetline,** made with water or retardant; **dozer line,** made by bulldozers.

crew: Usually a twenty-person team run by a crew boss. Two ten-person squads make up a crew.

crown fire: A fire spreading through the crowns of trees, often with enormous force.

dozer line: *See* **control line.**

ecosystem: A concept providing much of the philosophical foundation of contemporary environmentalism. In some dispute in scientific circles, the ecosystem paradigm holds that nature, if left "alone," will be "self-regulating" and achieve a "balance." In the popular culture, ecosystems have been identified in areas as small as a pond and as large as Planet Earth. Alston Chase's *Playing God in Yellowstone* contains a powerful critique of the ecosystem idea in the context of the Yellowstone area. *See also:* **natural regulation; Greater Yellowstone Ecosystem.**

engine: Any ground vehicle with pumping, water, and hose capacity.

fine fuels: Fuels such as dry grass, leaves, pine needles, and some kinds of undergrowth that are easy to ignite and burn fast. Also called **flash fuels.**

fire management officer (FMO): Principal officer on a district or administrative unit with fire responsibilities.

fire shelter: Life-saving device in shape of a floorless pup tent, usually carried in a pouch at belt.

firestorm: Extremely intense fire, usually generating crown fires, turbulent winds, firewhirls, spot fires, and unpredictable fire behavior.

firewhirls: A spinning, moving column of air within the firestorm that carries aloft smoke, debris, and flame.

fuels: Anything that burns.

Greater Yellowstone Ecosystem: The 8- to 18-million-acre area surrounding (and including) Yellowstone National Park. Also called the **Greater Yellowstone Area.** Environmentalists think that Greater Yellowstone—made up of Park Service, Forest Service, other federal holdings, and private lands—is a self-regulating natural body that would benefit from expanded preservation. *See also:* **ecosystem; natural regulation.**

hand crews: Firefighters organized and trained principally for operational assignments such as clearing brush and creating lines.

handline: Fire line made with hand tools.

helitack: Use of helicopters and trained airborne teams to achieve control of fires.

hotshots: Nickname for professional, year-round fire-suppression crews.

incident: A fire or other event, either human-caused or natural, that requires action by emergency-service personnel.

Incident Commander (IC): The individual responsible for the management of all incident operations.

Incident Command System (ICS): The system designed to allow the timely combination of resources, communications, and smooth operations during emergencies such as fires, earthquakes, or other natural or human-caused disasters.

let-burn: Shorthand for the policy of prescribed natural fire. *See also:* **prescribed natural fire.**

light on the land, light hand on the land: Firefighting tactic thought to cause least suppression damage, also called minimum-impact suppression; the terms caused confusion and dissent among firefighters in Yellowstone in 1988. Generally taken to mean an emphasis on confinement strategies and the use of hand tools.

236 **multiple use:** Guiding ethos of the U.S. Forest Service, generally taken to mean the mixed use of national forests,
allowing for a combination of timber harvesting, range-
land grazing, oil and mineral extraction, hunting, recreational activities, as well as the partition of preserved
"wilderness areas" within some forests. See, in contrast,
preservation.

natural fire: *See:* **prescribed natural fire.**

natural regulation: The idea, powerful in wildland management and environmental circles, that an "ecosystem," if
not interfered with by humans, essentially will regulate
itself, maintaining a "balance of nature." Natural fire is a
part of the natural regulation idea. *See also*: **ecosystem;
Greater Yellowstone Ecosystem; prescribed natural
fire; preservation.**

overhead team: Personnel assigned to supervisory positions
in the Incident Command structure, including the incident commander and his staff. Also called Incident Management Team. Referred to by line firefighters as
"overfed."

prescribed natural fire: Also called **prescribed natural
burns, natural fire,** and **let-burn,** prescribed natural fire
is the policy of allowing lightning-caused fires to burn
unless or until they threaten lives or structures. The natural fire policy—part of the Park Service's attempts to preserve the natural role of the elements within park boundaries—was at the center of the controversy over the Yellowstone fires. *See also:* **ecosystem; Greater Yellowstone Ecosystem; natural regulation; preservation.**

preservation: Guiding ethos of the National Park Service,
generally taken to mean that the human hand should be
as far removed from the ecosystem as possible and that
nature should be allowed to take its course. Sometimes
seen as detrimental to the Park Service's mission to preserve the parks "for the benefit and enjoyment of the people." In terms of fire, the Park Service is oriented toward
preserving the natural role of fire in the ecosystem; the
Forest Service, on the other hand, traditionally has been

oriented toward fire suppression. *See also:* **multiple use; natural regulation.**

resources: All personnel and equipment available, or potentially available, for assignment to an incident.

safety first: Primary law of wildland firefighting.

snag: Standing dead tree.

spot fires, spotting: Fires outside the perimeter of the main fires, usually ignited by wind-borne embers.

wildfire: An unwanted wildland fire requiring suppression action.

Sources: *Fireline Handbook,* National Wildfire Coordinating Group; *Fire on the Rim,* Stephen J. Pyne; *Greater Yellowstone Area Fire Situation,* Greater Yellowstone Coordinating Committee; *Incident Command System,* Oklahoma State University; *Playing God in Yellowstone,* Alston Chase.

Notes to Chapters
and Sources

Unless otherwise noted, interviews were conducted by the author and documents were obtained from the archives of the Public Documents Center, Mammoth Hot Springs, Yellowstone National Park.

Chapter I: Clover

The main sources for this chapter were interviews with Yellowstone National Park Chief Ranger Dan Sholly, pilot Curt Wainwright, and National Park Service Rangers Steve Frye, Phil Perkins, and Dick Bahr; *Guardians of Yellowstone,* by Dan Sholly with Steven M. Newman (William Morrow, 1991); and the official Incident Review Board Report of the July 14, 1988, Fire Shelter Deployment on the Clover fire, provided by an anonymous source.

Additional Interviews: YNP Superintendent Robert Barbee; NPS Rangers Randy King and Joe Fowler; YNP Chief of Public Affairs Joan Anzelmo.

Additional Documents: Clover Fire Situation Analysis, July 14, 1988; Clover Fire Escaped Fire Situation Analysis, July 16, 1988; Phase Two Final Report, Greater Yellowstone Area Fire Situation 1988, by the Greater Yellowstone Coordinating Committee, Chairman NPS Deputy Director Jack Neckels (hereafter, "Neckels Report"); Clover-Mist Fire Management Review, May 16, 1989; *History of the Greater Yellowstone Area Fires of 1988,* unpublished oral history of fires prepared by David W. Peterson for the U.S. Forest Service, contract number 40-84M8-8-1482 (hereafter, "Oral History, Peterson").

240   *Additional Books: Summer of Fire*, photo-essay by *Denver Post* columnist Jim Carrier, photos by Jeff Henry and Ted Wood (Gibbs Smith, 1989); *For Whom the Bell Tolls*, Ernest Hemingway (Scribners, 1940).

Chapter II: Mammoth Hot Springs

Important information was provided by Alston Chase's *Playing God in Yellowstone: The Destruction of America's First National Park* (Harcourt Brace Jovanovich, 1987); Stephen J. Pyne's *Fire in America: A Cultural History of Wildland and Rural Fire* (Princeton University Press, 1982); and Aubrey L. Haines' two-volume history of Yellowstone Park, *The Yellowstone Story* (Yellowstone Library and Museum Association, 1977). The author is particularly indebted to—and drew heavily on—Chase and Pyne for their insights into, respectively, contemporary environmentalism and the history of fire.

Interviews: Robert Barbee; Chief Naturalist George Robinson, YNP; Chief Biologist Don Despain, YNP; Ranger Lee H. Whittlesey, YNP; Thomas McNamee, Greater Yellowstone Coalition; Ed Lewis, Greater Yellowstone Coalition; Ralph Glidden; Sandi Robinson; anonymous.

Documents: Greater Yellowstone Coalition newsletters and annual reports, 1983 to 1992; "The Historical Roots of Our Ecologic Crisis," Lynn White, Jr., 1966.

Additional Books: Yellowstone Place Names, Lee H. Whittlesey (Montana Historical Society Press, 1988); *Exploring the Yellowstone High Country: A History of the Cooke City Area*, Ralph Glidden (Ralph Glidden, 1976).

Chapter III: Storm Creek

Primary sources included the document, "Storm Creek Fire from Beginning to End," by Custer National Forest, Beartooth District, Fire Management Officer George Weldon; related documents obtained from the Public Documents Center at Mammoth Hot Springs; and interviews with many local figures in the Cooke City, Clarks Fork Valley, and Red Lodge areas, including Mike and Beth Evans, Joan Hummiston, Hayes Kirby, Wade King, Pat Pierson, John Griscomb, and Dan Hogan.

Chapter IV: Mammoth Hot Springs

Forest Service officials often did not respond to requests for interviews or information. In some instances, then, the Forest Service side of the story had to be constructed from official documents, later newspaper and magazine stories, and David Peterson's "Oral History."

Interviews: Robert Barbee; Dan Sholly; Terry Danforth; Phil Perkins; Targhee National Forest official requesting anonymity.

Documents: Neckels Report; Fire Management Reviews, May 16, 1989; Oral History, Peterson; letter from John Burns to Robert Barbee.

Books: Fire in America, Pyne; *Introduction to Wildland Fire: Fire Management in the United States,* Pyne (John Wiley & Sons, 1984).

Chapter V: Snake River

By July, local press coverage of the fires was increasing. The author obtained extensive clippings from the *Denver Post, Billings Gazette, Bozeman Chronicle*—as well as several other smaller papers in Montana, Wyoming, and Idaho—and used them as background and backup throughout the book. Given the resources available, the local press did an outstanding job of covering the fires. Of particular note is the aggressive coverage of the *Billings Gazette* and—although not precisely a "local" news outlet—the reporting done by Jim Carrier and others for the *Denver Post.* Carrier's photo-essay book, *Summer of Fire,* provided important details. Other books of this genre included *Yellowstone and the Fires of Change,* by George Wuerthner; *Yellowstone on Fire!* by the staff of the *Billings Gazette;* and *The Fires of '88* by Ross W. Simpson. The Carrier and Simpson books were the best of this type.

Interviews: Dave Poncin; Robert Barbee; others.

Documents: Neckels Report; Snake River Fire Management Review; Shoshone Fire Escaped Fire Situation Analysis and related documents.

Chapter VI: North Fork

Interviews: Robert Barbee; Dan Sholly; Lorraine Mintzmyer; Jack Troyer; Dave Poncin; others.

Documents: Neckels Report; Oral History, Peterson; North Fork Fire Management Review; Fire Situation Analysis and other relevant documents. *Audubon* magazine, January 1989.

Chapter VII: Clover-Mist

This chapter was based on interviews with Curt Bates and other Clover-Mist firefighters; documents obtained from Mammoth Hot Springs, including photocopied pages of Bates' diary; the Neckels Report and the Clover-Mist Fire Management Review.

Chapter VIII: Grant Village

Interviews: Dave Poncin; Robert Barbee; Terry Danforth; Joan Anzelmo; others.

Documents: Neckels Report; Snake River Fire Review; ICS 202 (Incident Objectives) forms and other relevant documents; *Audubon* magazine, January 1989.

Books: Summer of Fire, Carrier.

Other: Inferno at Yellowstone, videotape by Cinevision Inc., 1989.

Chapter IX: Old Faithful

Sporadic coverage by the establishment daily press, newsweeklies, and television networks began in late July. The author obtained and used newsclips, articles, and computer records of network coverage.

Interviews: Donald Hodel; Robert Barbee; Joan Anzelmo; George Robinson; others.

Documents: Oral History, Peterson.

Books: Summer of Fire, Carrier.

Other: Taped segment of Hodel press conference, in *Inferno at Yellowstone,* Cinevision Inc., 1989.

Chapter X: West Yellowstone

Interviews: Don Despain; Robert Barbee; Joan Anzelmo; Jack Troyer; Dave Poncin; Curt Bates; Forest Service officer requesting anonymity.

Documents: Neckels Report; Area Command Management Review; photocopied diary pages obtained from Public Documents Center; narrative provided to author.

Books: Summer of Fire, Carrier.

Chapter XI: North Fork

Interviews: Robert Barbee; Dan Sholly; Steve Frye; Dave Poncin; Ed Francis.

Documents: "Fire," Thomas Hackett, *New Yorker,* October 2, 1989; Neckels Report; North Fork Fire Management Review; Fire Narratives and related documents obtained from the Mammoth Hot Springs archives.

Books: Summer of Fire, Carrier.

Chapter XII: Into the Black

Interviews: Robert Barbee; Dan Sholly; Steve Frye; Phil Perkins; Terry Danforth; Dave Poncin; Curt Bates; Dan Hogan; Lee Whittlesey; Hayes Kirby; Dave Majors; John and Anne Graham; Wade King; John Griscomb; others.

Documents: Neckels Report; Oral History, Peterson; Escaped Fire Situation Analysis and related documents.

Chapter XIII: North Fork

Interviews: Dave Poncin; Jack Troyer; Dan Sholly.

Documents: Neckels Report; North Fork Fire Management Review; "Oral History," Peterson.

Chapter XIV: Silver Tip

Some interviews for this chapter were conducted by research associate Dorothy Houser, who also obtained the Storm Creek Fire Shelter Deployment Report for the Silver Tip Ranch.

Interviews: Larry Sears; Steve Frye; anonymous.

Documents: Neckels Report; Storm Creek Fire Management Review; narrative of the Silver Tip Incident, by Larry Sears; Storm Creek Fire Shelter Deployment Report for the Silver Tip Ranch, including testimony of participants; Fire Situation Report #1 and related documents.

Books: The Yellowstone Story, Aubrey L. Haines.

Chapter XV: Bozeman

This chapter is primarily based on a videotape, obtained by the author, of the Area Command strategy meeting in Bozeman.

Interviews: Robert Barbee; Joan Anzelmo; Jack Troyer; Dave Poncin.

Documents: Neckels Report; Area Command Management Review; delegations of authority.

Chapter XVI: Storm Creek

Interviews: David Liebersbach; Hayes Kirby; Joan Hummiston; John and Anne Graham; Dave Majors; Dan Sholly; Joan Anzelmo.

Documents: Neckels Report, Storm Creek Fire Management Review; "While Yellowstone Burned," Micah Morrison, *American Spectator,* November 1988.

Books: Guardians of Yellowstone, Dan Sholly; *Exploring the Yellowstone High Country,* Ralph Glidden.

Chapter XVII: Old Faithful

A vivid portion of the videotape, *Inferno at Yellowstone,* records the scene at Old Faithful.

Interviews: Lee Whittlesey; Amy Vanderbilt; anonymous.

Documents: Numerous news clips; Neckels Report; North Fork Fire Management Review.

Books: Summer of Fire, Carrier.

Chapter XVIII: Clover-Mist

In this chapter—and only in this chapter—names were changed and geographic locations were slightly blurred to protect the privacy of sources.

Interviews: John Griscomb; David Anderson; Mike and Beth Evans; David and Kathy Hannahs.

Documents: Clover-Mist Fire Management Review; Storm Creek Fire Management Review; Neckels Report; Fire Behavior Evaluation #2 and related documents.

Chapter XIX: Mammoth Hot Springs

Interviews: Robert Barbee; Dan Sholly; Joan Anzelmo; George Robinson; Sandi Robinson; Phil Perkins; Steve Frye; others.

Documents: Neckels Report; North Fork Fire Management Review; Fire Behavior Evaluation #3.

Books: Guardians of Yellowstone, Dan Sholly; *Summer of Fire,* Jim Carrier.

Epilogue

Interviews: Robert Barbee; Dan Sholly; Don Despain; Phil Perkins; Joan Anzelmo; Dave Poncin; Park Service and Forest Service officials requesting anonymity.

Documents: Neckels Report; all Fire Management Reviews; Yellowstone National Park, Wildland Fire Management Plan, March 1992; news clips and magazine articles.

Index